AF821805

INVERTEBRATES FOR EXHIBITION

INSECTS, ARACHNIDS, AND OTHER INVERTEBRATES SUITABLE FOR DISPLAY IN CLASSROOMS, MUSEUMS, AND INSECT ZOOS

ORIN MCMONIGLE

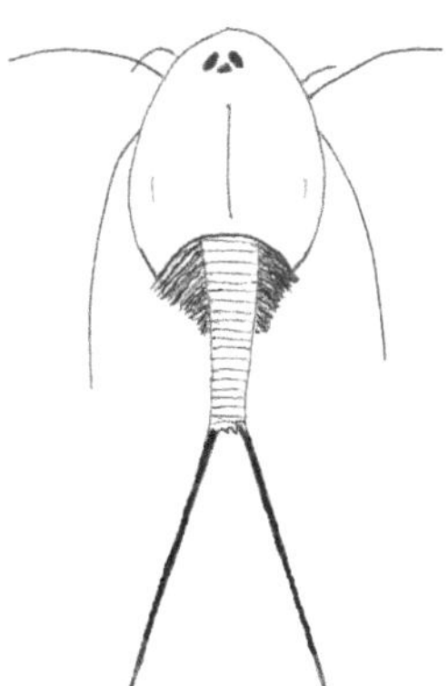

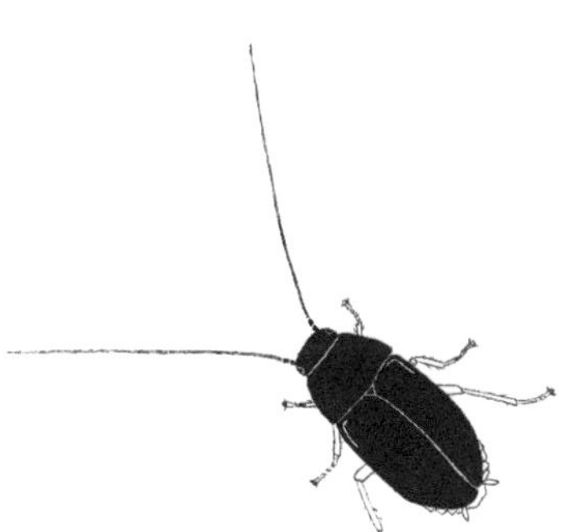

COACHWHIP PUBLICATIONS
CoachwhipBooks.com

Acknowledgements

Additional information thanks to Annie Lastar, Dr. Thomas Carefoot, Tom Larsen, Dr. Richard Hoffman, Dr. Luis M. Roth, Dan McMonigle, Jen-Pan, Peter Clausen, Mike Pickut, George Beccaloni, Dr. Aaron Dossey, Dr. Harley Rose, Richie Willis, Dan Marino, Steve Boddy, David Anderson, Christian Schwarz, Pernille Aalvik, Damon Debee, Don Ahart, Kelly Swift, and Steven Lenaerts. I appreciate the support from my wife, Sylvia, and my daughters, Kree and Gwynevere. I thank God for creating such an amazing array of invertebrates and giving us the desire to observe them.

Legal Notice

Please note that the possession, exhibition, purchase, or sale of certain native or exotic invertebrates may be illegal or require permits in some states and other localities. Please check applicable codes and state departments (wildlife, agriculture, etc.). A number of invertebrates in this book were photographed by permission at permitted institutions.

Invertebrates for Exhibition, © 2011, 2012 Orin McMonigle
Coachwhip Publications (Landisville, Pennsylvania), CoachwhipBooks.com

Front Cover: Death's Head Cockroach
Back Cover: (Top) Vinegaroon, Immature
(Middle) Emperor Scorpion
(Bottom) *Scolopendra heros heros*

McMonigle, Orin André.
Invertebrates for exhibition : insects , arachnids , and other invertebrates suitable for display in classrooms , museums , and insect zoos / Orin McMonigle.
Includes bibliographical references.

ISBN 978-1-61646-147-8

1. Invertebrates—Collection and preservation. 2. Insects —Collection and preservation. 3. Spiders —Collection and preservation. 4. Arachnida —Collection and preservation. 5. Crustacea –Collection and preservation. 6. Natural history —Technique. I. Title.

QL362.8 .M33 2012
579 —dc22

CONTENTS

MACLEAY'S SPECTRE, *EXTATOSOMA TIARATUM*

HARLEQUIN FLOWER BEETLE, *GYMNETIS CASEYI*

INTRODUCTION

This text provides information on caging and captive husbandry for a variety of invertebrates that can be displayed in exhibits, at bug shows, or in classrooms. The majority of species are readily acquired with little effort. Some large, impressive display animals are available almost anywhere, including *Bombyx* caterpillars and *Zophobas* beetles. A few can't be acquired without a permit (such as the exotic phasmids), can only be caught in certain areas of the country (like grasshoppers and some beetles), or are restricted according to species or genus in certain states (as with crayfish). The majority of species are easily maintained and many can be reared successfully in captivity. In nearly all cases hardy species were chosen, though hardiness assumes correct food and caging. Breeding requirements and developmental stages are detailed whenever possible.

Each species or group description begins with a short section titled "display." This offers helpful details relative to live exhibits and may be similar to or run contrary to some husbandry advice. For example, a scorpion display should show off the animal and make it visible to the onlooker—so it should have shallow substrate and be void of obstructive hides, even though the scorpion may exhibit burrowing behavior in its natural habitat. Another example is that various scarabs and stag beetles live healthy lives on shallow substrate, but cannot be successfully bred without the substrate depth required for egg laying. In some cases display animals can be cycled between the exhibit and the holding or breeding cage, while in most cases display conditions have no negative effect on reproduction, longevity, or health.

Cage decorations should often be smaller than the animals they accompany. Logs, rocks, and large pieces of wood are not needed to make invertebrates "feel at home." Many captive animals have been smashed as they dig under objects provided by well-meaning keepers. Likewise, if an animal seeks refuge under a rock for molting, a cage mate can destabilize the decoration, crushing the immobile specimen. Rhinoceros and flower beetles tarsi get stuck in logs, which tears off this part of their legs. Old beetle specimens in cages with logs often have no tarsi at all, but would not have lost body parts if correctly housed. Large decorations don't always kill or damage the subject—they may only obstruct the public's view and remove all purpose associated with the display.

While there may be examples of burrowing animals that require substrate to meet physical needs, that's not the case for species presently in culture. Burrowing animals in general don't need to burrow as long as moisture needs are met. It is important to understand that it's difficult to provide natural substrate conditions in captivity. Offering deep substrate to most millipedes, some beetles, and most notably the rhinoceros cockroach, *Macropanesthia rhinoceros,* is the direct cause of vast numbers of molting deaths in captivity. In nature they create caverns underground which can't be adequately supported in captive habitats or are collapsed by cage mates. The substrate caves in on them and the upcoming molt results in massive and mortal deformities.

It's a poor idea to transpose human comprehension, desire, and emotion onto invertebrates. There is no evidence to suggest a centipede can't feel as much "joy" digging through half an inch of sand as half a foot. Invertebrates tend to be just as "happy" with a glass hide or no hide spot at all.

Sections providing background, husbandry and life cycle information follow the display section. These sections are divided differently depending on the information appropriate to or available for the specific animal.

Red-Eyed Devil Katydid, *Neobarrettia spinosa*, Molting into an Adult

ORTHOPTERANS

Eastern Lubber Grasshopper
Romalea microptera

Display: 10- to 30-gallon aquarium with metal screen lid—ventilation is a necessity. Half a dozen specimens per 10 gallons. Foodplants shouldn't be so numerous as to block the view.

Background: The word lubber means a heavy, clumsy, or awkward fellow. These are all good words to describe the eastern lubber grasshopper, one of the heaviest of the United States' insects and one of the heavyweights of the grasshopper world. A large female measured from the head to the tip of the abdomen can be more than 80mm. (The wings are shorter than the body and add no length.) Lubber grasshoppers cannot fly, are slow-moving and have limited jumping ability. Their clumsy and awkward demeanor makes them an endearing and easily handled pet. Other large grasshoppers tend to have advanced fight and flight skills which would make it necessary to muzzle them and tear off their wings and back legs to turn them into unaggressive captives.

Though probably not the original reason for the name, the word *lubber* also means *a clown* and is descriptive of the flamboyantly colored outfit of *R. microptera*. Spots and stripes of red, yellow, black and orange paint the armor suit. The species is found across a huge area of the U.S., from North Carolina south to the tip of Florida and west through Texas. As is common for invertebrates with a wide range, there are a number of geographic size differences as well as color variations. The two primary adult color forms are usually referred to as 'black' and 'olive.' The black form has yellow edging which resembles nymph coloration and looks similar to the related, southwestern *Romalea (Taeniopoda) eques* (horse lubber grasshopper), but lacks the yellow vein patterns on the rear femur and tegmina. *Romalea microptera* color forms can interbreed in captivity, but otherwise do not throw nymphs of the other color. Nymphs of both forms are black, but black form nymphs have red edging after the first molt, while olive morph nymphs retain yellow edging through the subadult stage. There are many different geographic colors, including an unattractive intermediate of black and olive and an attractive pastel orange variation of the olive form.

Pair of Eastern Lubbers, Olive Form

Taxonomy: The eastern lubber is usually listed as the namesake of its own family, the Romaleidae, which includes a number of primarily New World and often large, colorful species that have two immovable spines at the end of the tibia. The older taxonomic scheme lists the Romaleinae as a subfamily of the Acrididae. This was the only species in the genus *Romalea*, although some texts mention "*R. guttata,*" which is a synonym. The closest relatives were species in the genus *Taeniopoda,* which are differentiated because the males can fly (seemingly only in certain wild situations, since they don't fly in captivity). Recently *Taeniopoda* was changed to a subgenus of *Romalea*.

The species name *microptera* means *tiny wing*. The wings are short compared to flying grasshoppers but *R. microptera*'s wings are incredibly large compared to the tiny round wings of the related *Brachystola magna* (plains lubber grasshopper). The hind wings are dark pink and black and can be raised up in warning. Forewings normally stretch halfway down the abdomen. The length of the tegmina can be longer in some populations and also varies according to how full the abdomen is (since wing length is expressed in relation to the abdomen).

PLAINS LUBBER, *BRACHYSTOLA MAGNA*

Food: Romalea's large size is misleading because it eats less than many grasshoppers 1/3 its size. It will nibble on a variety of leaves but very few can be used as a primary food. Some common plants such as privet, rose and blackberry are eventually eaten, but it seems only romaine lettuce is eaten with gusto. They seem to have almost no appetite when fed leaves other than romaine. Wheat or millet seeds can be planted in a shallow tray and used as food when the sprouts reach an inch or two. Dry dog food, cat food and carrots can be added to provide additional nutrients and to stretch out the primary food supply. Romaine lettuce can be used as the only food if preferred. Nymphs and adults have the same general diet, but fresh greens should be available daily for the first instar to ensure optimum survival. Only subadults and adults eat dry dog food pellets (most types)—either because they have more interest in protein or are less picky. Small nymphs prefer to feed on vegetation, but they go after the green 'vegetable' pellets in Beneful® brand dog food. (Palatability appears related to the vegetable content or color rather than the sturdiness of the pellet, because other far softer pellets in the Beneful® are refused.)

Behavior sets *Romalea* apart from the closely related horse lubber because it lacks a tendency towards cannibalism. Even when starved and overcrowded, it rarely bothers tank mates. Horse lubbers walk over and eat each other from behind, a rather ghastly behavior, since the victim appears quite unaware and walks around with half its body missing, even if it's just a head and two front legs. *Romalea microptera* also tends to grow larger than *R. eques* in captivity.

Romalea microptera can be considered a nuisance pest in gardens. Gardeners (ones that don't possess a love of beautiful invertebrates) tend to be frightened upon seeing a grasshopper the size of a small bird, even if there isn't any damage. Eastern lubbers are very slow-moving and harmless, occur in low numbers, and have large body mass, so pest control advice is to collect them and use a "smashing device." They are supposed to cause some damage to citrus and daylilies. In the lab they were offered orange leaves, which were not taken readily but were half-eaten after a few days in the cage. Daylily leaves were also offered but only nibbled on. Daylily flowers were half-eaten in a few hours but never finished.

Life cycle: Egg cases are laid two to three inches deep in the soil. The female buries her telescopic abdomen up to the back legs and forms a compact, foam-capped ootheca underground. Though they are glued together, a minimal amount of force shatters it into separate eggs. (It does not compare to the structured oothecae of mantids or roaches.) *Romalea* females make 3-5 oothecae, and each contains 25-40 eggs. Grasshoppers hatch after 4-5 months at room temperature. Unlike some types of grasshoppers, they do not molt upon reaching the surface. Hatchlings need some moisture and should be fed fresh greens every day or two. Nymphs molt every twelve to twenty days until the final molt, which can take more than a month. Adult females wait four to eight weeks after the final molt before laying eggs.

The following dates represent a single lab rearing and are provided as a timeline reference. Ova gestation temperature averaged 72°F. Nymphs and adults were kept around 78°F.

February 14	Eggs Laid
June 21	Hatching
July 4	First Molt
July 15	Second Molt
July 27	Third Molt
August 12	Fourth Molt

Husbandry: The most difficult part of the eastern lubber life cycle is gestation of the ova. Exposed, dry conditions wipe out every egg in as little as a few hours and hatching rates can be extremely low without a repeatable and dependable methodology. The best method requires no upkeep. A clear, 32oz deli-cup and lid with no air-holes is used. The container is filled 2/3 with inexpensive play sand mixed with 10% coconut fiber or peat. It should be very damp but not waterlogged. The surface should always stay damp enough to clump and not separate when pinched, or it is too dry. This is placed in the cage, without the lid. After a few oothecae are laid (which isn't always easy to determine) any frass is removed from the surface and the lid is placed on top with a date label. The surface moisture can be checked from time to time but really should not change. Hatchlings must be moved to a vented cage with food within a day or two to survive.

It is very important to provide angled surfaces, such as branches, so nymphs can molt successfully when kept in a smooth-walled container like an aquarium or plastic small animal cage. They will hang upside-down from the branch or lid to molt. Virtually defect-free molts can be achieved if they are kept in a screen-walled cage such as those designed for chameleons.

Good ventilation is very important, because stagnant, damp air will cause death in days to weeks. The lid, or at least one side, should be made of window screen. Unless the food or atmosphere is extremely dry, daily misting is unnecessary.

Eastern Lubber, Subadult

Remarks: This wonderful little clown entertains children at many insect zoos. I've seen them on exhibit at two of the oldest and largest U.S. insect zoos—the Smithsonian in Washington, D.C., and Insect World in Cincinnati, Ohio.

Romalea microptera is a wonderful and seemingly friendly display animal. Beyond excellent disposition, size, and color, it is easy to feed in the winter as there is no need to hunt down and collect food plants. (Eastern lubbers much prefer romaine lettuce to other foods.)

Horse Lubber Grasshopper
Romalea eques

Display: 10- to 30-gallon aquarium with metal screen lid—ventilation is a necessity. Half a dozen specimens per 10 gallons. Foodplants and molting branches shouldn't be so numerous as to block the view.

Background: The horse lubber grasshopper isn't large enough to saddle, but it is one of the largest North American orthopterans. The species name *eques* comes from the Latin word for *horse* and gives rise to the common name. It's difficult to say if the species name refers to the massive size, the horse-like face or if there was some other association with horses. It ranges from Arizona to Texas in the United States, and is found south as far as Costa Rica. There are presently five other *Taeniopoda* species found in Mexico and farther south, including *T. auricornis, T. burmeisteri, T. centurio, T. picticornis* and *T. superba.* Considering the minimal differences, it's not surprising the genus *Taeniopoda* was recently reclassified as a subgenus of *Romalea*. *Taeniopoda* still exists—as the subgenus—but the binomial is now *Romalea eques*. *Romalea eques* is a member of the tribe Romaleini in the lubber grasshopper family Romaleidae. The Romaleidae is small in number but contains many of the largest grasshoppers, including the largest of all, the birdwings in the genus *Tropidacris*.

HORSE LUBBER, MALE

Horse lubbers are gaudily colored and one of the few insects to employ every one of the usual warning colors—black, red, white, yellow and orange. Nymphs are black with yellow and white margins. Adults are black with orange margins. Contrasting red and black underwings are flashed in warning if large predators (such as humans) approach. The wing flashing is common in the wild but seldom practiced in captivity. Birds, mammals and even toads won't eat them. Predatory invertebrates like tarantulas, centipedes and mantids feed readily on lubbers—they apparently don't mind the taste and probably can't see colors anyway.

Life Cycle: Females deposit eggcases (containing around forty eggs) 3-5 inches deep in damp, sandy soil. They retain eggs while searching for a good area, but will eventually deposit eggs even on the floor of a glass-bottom cage. It is important to provide egg laying substrate in time because exposed eggs desiccate rapidly and other grasshoppers in the cage will chew on or devour exposed eggs. Some foam is formed around the eggs, but primarily as a cap, so it is a loose mass and only well-protected below the surface. As long as the eggs are never allowed to dry out (deep substrate may become dry without drying eggs) they hatch after 6-12 months at room temperature. For ova survival, sandy substrate should always be kept moist, but not clogged or saturated with water.

Romalea eques emerge two to six at a time for a week or two—not as a single group, as seen with *R. microptera*. They slowly worm their way to the surface, but eventually die if they can't break free. The eggs should not be reoriented since the hatchlings tend to go straight away from the top of the egg and can be doomed if the egg is facing down. Avoid digging up eggs. Those that hatch against glass or clear plastic often do a few loops resulting from a loss of direction due to the light, but usually make it to the surface anyway. Their first molt begins when they are no longer under compression. They do not molt

Horse Lubbers Hatching

against the glass under substrate (proving light is not the factor that instigates the molt) but will molt in the dark if they end up in an underground cavern in the sand. Prior to hatching, make sure to tamp down sand to prevent cavities large enough to trap emerging young. The molt begins within minutes of breaking free. Compression will postpone molting for up to a few hours, but they will die in the old exoskeleton if they don't molt in time. The wormy hatchlings have full-length curvy legs, but cannot walk until they undergo the first molt. While eggs can be kept in a permanently damp, barely ventilated container, nymphs must be moved to a screen or part screen cage within hours to survive.

Baby lubbers feed on leatherleaf viburnum, wandering Jew, raspberry, rose, etc., but require a supply of fresh romaine lettuce to do well. Old lettuce should be discarded and replaced daily. The survival rate is high only with a constant supply of fresh romaine. Any type of dry cat or dog food pellet is also eaten, but cannot be used as the main diet.

The post-hatching molt is the only one that can be successfully completed on a surface. Nymphs can molt successfully the subsequent one or two molts as they cling from a smooth surface. During later molts the nymphs must hang from a stick or screen side and will become deformed or stuck if they fall or molt on the bottom of the cage. If an aquarium is chosen over a screen cage, place it on the end or side so the screen lid becomes a wall. A branch stretching from the bottom corner to the top corner is helpful. Each molt requires two to three weeks of feeding and there are about half a dozen molts in total. Molting surfaces become more and more important and molts grow more dangerous as they get closer to maturity, because the animals are heavier.

Horse Lubber Hatchlings

Horse lubbers develop wings on the molt to adulthood. Adults are thick and sturdy and can exceed two and a half inches in length. The flightless females are bulkier and the abdomen sticks out past the wings. Males can also have short wings, but more often have wings that exceed the length of the abdomen. Males with longer wings are capable of short flights. Both sexes live three to four months after reaching adulthood.

This species is extremely cannibalistic and famous for its tendency to accumulate on highways. They are attracted to their own dead as a food source and in turn are run over while feeding, which continually adds to the pile of death. If they're very hungry, they just walk over and eat each other, starting from the rear and often attacking in a ravenous but terribly slow-moving group. This leaves the victim only a sad little head and front legs with which to walk around. Every day or so a new victim is singled out and consumed by the group. Fortunately it's not necessary to keep them separate. As long as fresh romaine lettuce is regularly included in the diet the problem is nearly eliminated. Of course, for a novice keeper, keeping a group can be rather disconcerting.

HORSE LUBBER, LATE INSTAR NYMPH

Remarks: Romalea eques is a beautiful, Halloween-colored species with one of the sharpest patterns of warning coloration for any grasshopper. The fiery yellow and orange margins and antennae set off the black body. The underbody looks like it is made from a piece of shiny black metal. The horse lubber is one of the least difficult orthopterans to rear, with nearly the same level of ease as *Romalea microptera.*

STICK GRASSHOPPER
PROSCOPIA SCABRA

Display: 10- to 30-gallon aquarium with metal screen lid—ventilation is a necessity. Half a dozen specimens per 10 gallons. Specimens blend in with twigs, so foodplant should be chosen so that specimens are easy to see.

Background: The Proscopidae are a peculiar family of grasshoppers found only in Central America and northern South America. Members of this group bear a striking resemblance to stick-like phasmids, except for the unique face. As in other grasshoppers, the rear legs are adapted for jumping. Nymphs and adult males can jump twelve inches horizontally. Mature females lack this ability. Adult males are three inches long and striped in bright aqua-blue and dark gray. Female *P. scabra* are large, brown animals averaging six inches in length.

Husbandry: Proscopia scabra are not omnivorous like the majority of grasshoppers, and after hatching are kept more like a phasmid than a

STICK GRASSHOPPER, MALE

STICK GRASSHOPPERS, PAIR

grasshopper. They feed exclusively on fresh leaves of select plants, including the usual phasmid staples: rose, blackberry, and raspberry. A branch of foodplant should be placed in water to keep leaves fresh. The water container should have a lid with a punched hole or paper towel to prevent the animals from climbing inside and drowning. Moreover, the container dimensions should be large enough to keep the cutting supplied for days and should be wide and heavy enough not to fall over.

An aquarium with 4″ of damp sand mixed with dirt and a screen lid makes a good cage. A container of substrate for oviposition will be necessary if a screen cage is used, but females may have a difficult time finding a container of substrate. Eggs laid on the floor of the cage dry out within minutes. Even if encountered and buried immediately by the keeper, the survival rate of eggs laid in the open is minimal.

The female drives her entire abdomen deep into the soil to lay an eggcase consisting of thirty to fifty eggs. The abdomen doesn't telescope during oviposition as with most grasshoppers. However, her abdomen is very long, which allows her to bury the eggcase three to four inches deep. Ova require three to four months to hatch and the substrate must be kept damp the entire time. Conversely, excessive dampness can suffocate the ova. Nymphs begin their first molts as soon as they reach the surface after hatching.

RED-EYED DEVIL KATYDID
NEOBARRETTIA SPINOSA

Display: 5- to 10-gallon aquarium with metal screen lid—ventilation is a necessity. One specimen per cage. Introduce only a few prey at a time.

Background: This vicious monster is a sought-after exhibit specimen and possibly our most spectacular native orthopteran. The eyes are red while the body is bright green and white. Also known as the greater arid land katydid, *N. spinosa* is an aggressive, obligate predator. The gigantic, spiny legs make it appear even larger than it is.

The powerful jaws are readily used in defense, and the limber body makes it nearly impossible to hold a specimen in such a way that it can't bite. Its defensive display includes a reared-up body, open jaws and spread wings. The rear (flying) wings used in the deimatic display are black with blue spots and are too short for use in flight.

RED-EYED DEVIL KATYDID, IMMATURE

Reproduction: The mature female has an immense, sword-like ovipositor and a larger abdomen than the male. (The head, legs and thorax of the two genders average the same size.) The ovipositor is short and difficult to see on hatchlings, but enlarges with each molt and is obvious by the third instar. Male nymphs look similar but have no ovipositor and upon maturity the outer wings (tegmina) look rather different, since they are developed for stridulation.

The adult male has an impressively loud mating call which is sung during the day in captivity. Although both genders are predators, the primary risk of cannibalism is during growth and molting. If given space for the introduction, losses during mating attempts are rare.

RED-EYED DEVIL KATYDID THREAT DISPLAY

Females drive the ovipositor deep into the substrate and usually lay two to ten eggs each day. Damp, mostly sand substrate can be placed in a 32oz deli-cup for oviposition. Eggs are tan cylinders about a 1/3″ long and 1/16″ wide. The greatest barrier to successful reproduction of this species is hatching the eggs. If kept above 85°F they can hatch in three months, but in my experience it took more than two years with alternating hot and cold cycles to finally get them to hatch. A second attempt was no quicker. Ova should be kept damp the entire time, but resist some drying. Fertile eggs

RED-EYED DEVIL KATYDID EGGS (1 YEAR OLD)

contain clear liquid during most of gestation. Nymphs can only be seen weeks before hatching, so opening an egg to determine fertility of eggs laid by the same female does not work. Healthy eggs will remain plump, while those that die or were never fertile deflate and begin to disintegrate.

Nymphs: Upon hatching, the small nymphs are hungry, but are too weak to catch crickets. On day one they should be given a small piece of banana to eat, which gives them the energy to hunt. The next day they can be given a cricket sized at 1/4 to 1/2 their mass. A primary diet of crickets works well. In captivity, all stages catch and eat crickets and cockroaches, but will also chew on dog food pellets, banana, and some other foods. For successful development, the diet should consist almost entirely of live food.

Molting is tricky. Small specimens usually have no trouble. But, as they grow larger, they still need to hang upside down to molt, and their great weight makes them fall from small or smooth surfaces. For molting, the lid should be metal window screen (fiberglass screen can be chewed through) and at least 4″ across.

PHASMIDS

TWO-STRIPED WALKINGSTICK
"SKELETON STICK" FORM
ANISOMORPHA BUPRESTOIDES

Display: 5- to 10-gallon aquarium with metal screen lid—ventilation is a necessity but consider a hood setup to block defensive spray. Half a dozen specimens per 2 gallons. Foodplants shouldn't be so numerous as to block the view.

Background: Anisomorpha buprestoides is one of the most common and widespread stick-insects in the U.S. It is a handsome phasmid which is sometimes brightly colored. This contrasts to most phasmid species that are often the same color as sticks or leaves. Most *A. buprestoides* are very dark brown with two brown to light tan stripes running the length of the body, including the head. I remember a picture in a book showing several pairs on a palm leaf and thinking "Wow! ...Too bad they don't really look like that." While the common brown form varies somewhat in color intensity across its range from Texas to Florida, I thought this picture was simply a result of overexposure or overzealous photo editing.

The "Skeleton Stick" is a descriptive name for a geographic color form of *A. buprestoides* that occurs famously in Ocala National Forest. The body is jet black on top with a column of bone-white, segmented markings extending down the middle. The color pattern very much resembles a skeleton costume. The legs are jet black, the

TWO-STRIPED WALKINGSTICK, "SKELETON STICK" FORM

underside chalky white and the antenna segments are black with white margins. It is very common in its locale. It is not a new discovery; Hetrick reported first observing them in 1947 at Juniper Springs, Marion County, Florida.

Anisomorpha are often warningly colored, as they have large glands producing a noxious defensive secretion. The secretion contains a unique volatile chemical called anisomorphal (of course), which causes sneezing and watery eyes within a foot or so. Dogs have been reported to sustain permanent eye damage. However, they may neither understand the need nor have the ability to rinse with water. They may damage their own eyes trying to wipe the burning away.

At adulthood, females of this species are from 60-80mm while the males seldom grow to more than 35mm. The Skeleton Stick is in the medium to large size range for the species.

Two-Striped Walkingstick, Normal Form

Mating: Monogamy is a hallmark of *Anisomorpha.* The tiny male latches onto the female and usually doesn't let go till death. The male can latch onto a female up to two molts before she reaches adulthood and only temporarily releases when she needs to molt. He skillfully climbs around on her while she is molting and doesn't disrupt her.

Ova: Females lay up to 300 eggs over a five-month period. The shell is mottled gray but turns brown with age. 1/2″ nymphs hatch after three to four months at room temperature. As long as ova are

Two-Striped Walkingstick Eggs

not allowed to dry out the hatch rate is high, 80-90%. Eggs should be kept damp at all times.

A few inches of substrate is useful to maintain some humidity, but is not required if the eggs are removed and placed in a container for hatching. High humidity is preferred but medium humidity is acceptable as long as ova do not dry out.

Food: Anisomorpha accept fresh leaves of oak, rose, blackberry, strawberry, pyracantha, privet, leatherleaf or pragnese viburnum, Chinese privet, *Ligustrum sinense* (both green "northern" and green and white "variegated" varieties) and many commonly used phasmid foodplants.

Remarks: The Skeleton Stick ranks among the most beautifully colored phasmids in the world. It is even more vivid in person than in photographs. *Anisomorpha* are among the easiest phasmids to culture and combined with unbelievable coloration should make this an extremely popular display specimen.

Macleay's Spectre
Extatosoma tiaratum

Display: 10- to 30-gallon aquarium with metal screen lid, or full screen cage. Six specimens per

MACLEAY'S SPECTRE

10 gallons. Foodplants shouldn't block the view, and should contrast with the animals.

Background: This cryptic beauty from Australia has been cultured since at least the 1960s and is a must-have for the average insect zoo. It is easy to see in most cages, amazes observers, and is easy to care for. Females are huge, finger-thick creatures that hold the spiny abdomen in an arch, which commonly generates "scorpion" comments from observers. Many cultures contain no males and reproduce through parthenogenesis, but males are handsome creatures in their own right. They are scrawny and have no dorsal abdominal spines but are outfitted with huge flying wings and can take off in a moment.

Life Cycle and Husbandry: Females catapult eggs—a fascinating "egg laying" method—which can be heard now and then as the eggs bounce off the sides of the cage. Average, well-fed females produce 300 ova, but much higher numbers have been reported. The egg shell isn't easily damaged and is not prone to desiccation. Ova can be kept damp and will hatch in around five months, or can be kept dry and later moistened to postpone hatching a few months. Eggs from stocks that are bisexual (both males and females present) or parthenogenetic have high hatch rates.

Skinny, black nymphs with reddish heads hatch out from golden-brown mottled eggs and run ceaselessly around the cage. The frantic behavior and coloration resembles ants. Within a day or two the hatchlings settle down on foodplants, cease daytime movement, and transform to a brown-tan color after they've begun feeding. Spines running down the back of the abdomen make it easy to differentiate females following the first molt. Maturity is reached in about four months and adults usually live four to six months after that.

Extatosoma species accept fresh leaves of oak, rose, blackberry, strawberry, pyracantha, privet, mulberry, leatherleaf or pragnese viburnum, Chinese privet, and others. Pyracantha (firethorn) is recommended in northern areas since it is prolific, fully evergreen, and accepted readily by hatchlings. Cuttings should be kept fresh in water. Water should be covered, though only frantic hatchlings are prone to drowning themselves.

Texas Unicorn Mantis

Brunner's Mantis

MANTIDS

Texas Unicorn Mantis
Phyllovates chlorophaea

Display: 1- to 2-gallon aquarium or plastic jar with metal screen lid—ventilation is a necessity. Cohabitation is possible in large cages but one specimen per cage is recommended. Prey items should be eaten within 30 minutes or removed.

Background: The Texas unicorn mantis for years had been one of the most sought-after species, as it was noted in a number of books, but livestock was never available. All present specimens (even those cultured in Europe) come from oothecae laid in January 2007 by a single female captured by a zookeeper in Brownsville, Texas. This species is known to occur south all the way to northern South America and there are a number of similar looking relatives found across the tropical New World. The only related species in the U.S., the Arizona unicorn mantis, *Pseudovates arizonae*, known only from Arizona, is similar, but is a little smaller and has small lobes on the legs. *Pseudovates* nymphs can be collected in winter in Arizona because the oothecae do not diapause and it's likely *Phyllovates* are also unable to diapause. Arizona unicorn mantis stock has been started half a dozen times since 1994, but fizzled out each time—both species have sturdy nymphs and are easy to rear, but the Arizona species is difficult to mate successfully in captivity.

Life Cycle: Nymphs are sturdy and the survival rate is extremely high with moderate care. They readily feed on fruit flies for the first few instars and later can be switched to small cockroaches or crickets. Nymphs should be fed every two to three days and misted on the off days. They can survive long periods between feedings, many weeks for older nymphs, but delayed feeding is not recommended. Growth is slow compared to seasonal mantids (commonly six to nine months rather than three) and is influenced primarily by quantity and frequency of feeding. Cannibalism occurs but nymphs can be successfully reared in groups with few losses. Although one of the hardiest species kept in captivity, it can be a difficult species to mate and produce fertile oothecae. Males are a little smaller and mature more quickly in captivity, but are usually in good health when females have matured. The problem is that males often show little interest in females. Keeping the males in a separate room and at temperatures above 80°F increases their interest. Females produce four to ten globular oothecae which are about a half-inch in diameter. Fertile oothecae hatch about forty days after being laid.

Brunner's Mantis
Brunneria borealis

Display: 1- to 5-gallon plastic creature cage or aquarium with metal screen lid—ventilation is a necessity. One specimen per cage. Prey should be removed if not eaten immediately.

Background: Brunneria borealis is a lengthy creature, half an inch longer than our longest mantid, the adventive Chinese mantis, *Tenodera sinensis*, and longer than the largest *Hierodula*. The adult female is brilliant, a slender mantis with thick antennae, tiny pink wings and a bright green body. A pink stripe flanked by two yellow stripes run the length of the abdomen. This adult mantis usually takes on a different pattern when captive-reared; the tiny wings are purple, the green body is pink and the yellow stripes are white.

Husbandry: Brunneria borealis is uniquely feminist. She is the only mantis on earth known to naturally reproduce through parthenogenesis—no male exists or is needed. Ootheca hatching is likewise unique. The normal hatch time at room temperature is 4-6 months, not the usual four to six weeks. Two to three dozen nymphs emerge on different days, up to two months apart. However, following a cool period of three to four months at 50°F, they hatch at room temperature in four to six weeks and can all hatch out in a few days.

Hatchlings are large but have proportionately small raptorial legs. These slender forelegs lend to the belief that *Brunneria* eats only very small prey. However, an adult female easily dispatches the largest adult cricket—even if she can't finish the entire meal. Hatchlings take time to begin feeding and die easily from desiccation. Damp substrate, good ventilation and daily misting are recommended. The primary challenge to keeping this species is preventing bad molts during the last few instars. Large surfaces are needed because of the very long body and legs. Tall deli-cups have too small a lid diameter for successful molting. Larger nymphs will do best in a cage with a 5" or greater diameter and height. A paper towel can be glued to the lid or side of the cage to improve molting surface area.

Mantis oothecae come in an amazing variety of shapes and sizes, but the basic structure is the same. Opposite the point of attachment, or opposite the back, with a hanging eggcase, there are rows of apertures for the eggs which allow the hatching nymphs to escape. On most eggcases, these apertures are very conspicuous because they're a different color, which makes it easy to tell the 'front' or point of emergence. On some oothecae like *Parasphendale* the apertures aren't a different color but they're easily seen at hatching time. Of course *B. borealis* has decided not to follow the basic structure and so she creates an ootheca where every hatching nymph must come out individually from a single aperture at the end of the "spine."

SHIELD MANTIS
RHOMBODERA STALLI

Display: 1- to 5-gallon cage with screen lid—ventilation is a necessity. One specimen per cage.

Background: This handsome mantis has only been in culture since 2007. The stock is from Java, Indonesia. Nymphs are usually green after the first molt, though some stay brown and do not change back to green until the final molt. The top of the abdomen and basal two-thirds of the flying wings are colored a brilliant pink. The bulky adults are among the larger mantids at 3-3.5". A similar species, *R. basalis*, has been kept in captivity longer, but is prone to difficulties and so is uncommon. *Rhombodera* is similar to the closely related *Hierodula*, with the only notable difference being the widened prothorax (the "shield").

SHIELD MANTIS, MALE

Shield Mantis with Eggcase

SHIELD MANTIS PAIR: THE HAZARD OF MATING

Life Cycle: The large globular oothecae are an inch across and terminate in a pointed tip. Oothecae should be kept in a sealed deli-cup. A drop of water can be added to the bottom of the cup. There should be no wood attached to the ootheca (or other materials that could mold). One to two hundred nymphs hatch out about five weeks later. These should be immediately moved to a cage with screen vents. The first few instars are usually kept together and fed fruit flies. Following a few molts they can be separated and fed small crickets or early-instar lobster roaches. They reach adulthood in three months with adequate feeding frequency. Adults are usually undersized if fed only crickets. Overall, this is a hardy and easy to maintain species, but the molt to adulthood can be problematic if the ambient humidity is low and subadults are not misted daily. The female usually produces her first ootheca well over a month after reaching maturity. Males that survive mating live three to four months, while females usually continue on for about six months.

GHOST MANTIS
PHYLLOCRANIA PARADOXA

Display: 1/2- to 30-gallon aquarium with screen lid—ventilation is a necessity. Large cages can house a number of specimens, though cannibalism may occur.

Background: Phyllocrania paradoxa hails from tropical Africa. The pronotum is shaped like an arrowhead. The oblong head is topped with a huge, leaf-like extension. The legs and abdomen are decorated with multiple leafy extensions. As usual, the male is less bulky than the female. However, his large flattened wings give him the appearance of being similar in size. This mantis is usually brown, but can be black, green, or yellow.

FEMALE GHOST MANTIS

Husbandry and Life Cycle: The smooth oothecae are thin and usually formed along fine branches. They are long and flat like most hymenopidid (flower mantid) oothecae, but possess an unusual, skinny, stick-like process about half an inch long, jutting off the end. This may improve camouflage in nature, as it doesn't affect hatching. Temperature should exceed 72°F, and humidity should be kept high to prevent desiccation.

Five to forty tiny babies emerge from the eggcase after about two months. The hatchling nymphs look like groups of small black ants. They are tiny, but feed well on *Drosophila*. Nymphs kept together cannibalize more or less relative to the amount of food offered, cage size, and number of perching areas.

Up to ten adults can be comfortably housed together in a 29-gallon, planted terrarium. Adults do not tend to eat each other and captive mating is effortless. Nevertheless, many of the eggs in the oothecae can be infertile or not develop, so that many produce only a handful of nymphs.

GHOST MANTIS, FEMALES
(COLOR VARIATIONS)

After a few generations in captivity the fertility level normally improves.

Cryptic Pretty Mantis
Sibylla pretiosa

Display: 1/2- to 1-gallon vented container for individuals, or 5- to 10-gallon aquarium with screen lid for up to a dozen. Cohabitation is common in larger cages, but cannibalism or damage can always occur.

Background: This handsome African species lives in western and southern savannahs. This and *S. dives* are the only savanna species, as all other Sibyllidae occur in rainforests. This may be the reason they are rather sturdy in generally dry captive conditions. All stages can accept temperatures from 80-95°F, and high temperatures are commonly suggested, but they develop perfectly well in the low 70°s.

The peculiar forked, heart-topped head armature is among the most intricate, possibly outdoing *Phyllocrania*. Nymphs are wispy with long legs that suggest a stick figure-mantis. Only at adulthood, with the addition of large wings, do they appear bulky. Adults females are two inches, males a little shorter.

Life Cycle: Oothecae hatch after about forty-five days and usually contain two to three dozen individuals. This species is normally housed communally, even as nymphs grow large, but if they are forced into close proximity by limited space and perching areas, cannibalism or lost walking legs are likely. Fortunately, the walking legs grow back readily, and are nearly full-size after one molt, though the leafy leg extension is not so easily or quickly regrown. They prefer active or flying prey that comes to them, and will only hunt down terrestrial prey such as lobster cockroaches when kept in bare-bottom, small cages. They accept fruit flies into the late instars and can be switched to house flies at adulthood. (Puparium are readily available from some suppliers.)

CRYPTIC PRETTY MANTIS, SUBADULT

When the adults mature, they have brown, drab wings. Unlike other mantids presently cultured, the wings do not stay the color they first harden into. After about two weeks the tegmina turn green. At this time they should be ready to mate, and captive mating is rarely a challenge. Each female produces around a dozen narrow oothecae from 1/4-1/2″ in length. These are also unique in color, as they don't fade to brown but remain bright lemon-yellow.

ROACHES

MADAGASCAN HISSING COCKROACH (HISSER)
GROMPHADORHINA PORTENTOSA

Display: 2- to 30-gallon cage with enough ventilation to prevent mold growth, relative to ambient humidity. Large numbers of specimens can be kept together depending on floorspace and hiding areas, but more than a dozen in a 5-gallon aquarium can lead to damaged and chewed up specimens. They hide during the day so a dark room with red lighting can be used, or cork bark slabs can be propped against the front of the habitat so that diurnal hiding areas are the viewing areas.

MADAGASCAR HISSING COCKROACH

Background: This is easily the most famous pet insect. The list of movies this creature has appeared in is longer than the list of most movie stars. Everyone is familiar with the hit show where people ate or were buried in hissers on their quest to win prizes.

It could be argued that hissers have helped promote insect keeping more than any other invertebrate; many insect keepers began their journey with this animal. On the flip side, not so long ago when these "new" pets were featured on a news station in Florida, a number of people armed with "cockroach hysteria" and lack of knowledge called their local officials in fear, which reportedly caused the Florida arthropod permit requirements. Hobbyists have been keeping *G. portentosa* since at least the mid '60s and to this day there are no confirmed feral populations. (Reports have turned out to be either the native *Cryptocercus punctulatus,* that looks very much like a small hisser, or *Eurycotis floridana*, another native, which doesn't look much like a hisser.) Even so, adventive *Gromphadorhina* would do no harm since they are neither structural nor agricultural pests.

Despite cockroach myths, hissers do not carry diseases. Some people are allergic to airborne dust particles from the shed exoskeletons. However, there are uncountable shed insect exoskeletons outdoors, so it would seem a person with this allergy would need a good air filter in the house either way. There is a myth that cockroaches cause asthma, but the actual study showed only that in the USA (similar studies in other countries showed no correlation) persons who already had asthma *and* cockroach allergies were more likely to have more severe asthma attacks in the presence of an infestation.

Hissers are great display animals and can be handled easily. *Blaberus* species are preferential for open-top displays since they can't climb the smooth walls of an aquarium or tub. If hissers are used, the presenter may spend the entire time pushing the animals back down the sides.

These animals hiss to ward off predators. The sound is made by air rushing through the spiracles as the abdomen is contracted. The hiss is a warning of unpalatability. If molested further, the roach can spit fluid and jab the attacker with strong spiny legs. The spines can pierce human skin, which gives rise to the bite myth. At the very worst, spines may draw a barely visible speck of blood. If the attacker isn't stopped by these defenses, the thick armor presents a further challenge.

The most common taxonomic scheme for hissing roaches is:

Order Blattodea
  Superfamily Blaberoidea
    Family Blaberidae
      Subfamily Oxyhaloinae
        Genus *Gromphadorhina*
        Genus *Elliptorhina*

The Blaberidae contains nearly every roach that gives birth to live young, which includes most of the common pet roaches like *Blaberus* and *Eublaberus*. The Old World subfamily Oxyhaloinae contains around fifteen genera, including all hissers and the familiar *Nauphoeta* (*N. cinerea,* lobster roach), *Rhypharobia* (*R. maderae,* deadleaf roach) and *Henschoutedenia* (*H. flexivitta*, giant lobster roach). The namesake genus of this subfamily is *Oxyhaloa,* which includes *O. buprestoides*, a fully-winged, metallic green and orange jewel beetle mimic. Possibly the only unique shared characteristic of members of this group is that the male produces both mating pheromones and an aphrodisiac substance.

"PRINCISIA" HISSER

FLAT HORNED HISSER (*AELUROPODA INSIGNIS*)

All described hisser species are found on the largest island in the world, Madagascar. Original descriptions were based primarily on the shape and relation of the males' horns, whereas present taxonomy focuses more on the male's genitalia. Except for the *Elliptorhina*, all of the following species have the same male genitalia. They readily interbreed and should never be kept in the same cages, even for transport to a display.

Hissers in the Hobby: Gromphadorhina portentosa (Madagascan hissing cockroach) is the most commonly kept member of the genus. *Gromphadorhina* means "snout horn", or more literally, *gromphado* = "like a sow's", *rhina* = "horn." *Portentosa* refers to the hiss, which is a portent or warning. Some culture stocks are labeled as *G. picea,* though in culture *G. picea* = *G. portentosa.* What are even more strange are the occasional cultures labeled *G. picea* that aren't even the black stock of *G. portentosa.* (*Picea* means "black," or more literally,

WIDE HORN HISSER

DWARF HISSER

"pitch black.") The pronotum is always black, but the mesothorax, metathorax and abdomen can be tan, orange, yellow, black, brown or red-brown.

This species has been in culture for many decades and distinct culture stocks exist. Usually stocks have been selected for fast, easy breeding, or color and size. For obvious reasons, fast breeding stocks are small and the most common. Nymphs of any stock can be black. However, only certain stocks produce entirely black adults. Males of common stocks range from 35mm-65mm, females a little longer. (Keep in mind, length measurements are extremely subjective because the abdomen expands and contracts.) Giant stock adults grow longer and are much more heavily built. The most visible differences in giant stock include a wide body, massive head, thick legs and a broad pronotum with horns that look more like horns than bumps.

Black and giant stocks produce some reversions, midgets and deformities that should be removed to strengthen the strains. Reversions in black stock are easy to single out. Small males of the giant stock can be picked out by the development of a ridge on the pronotum, however it is extremely difficult to distinguish an immature or tiny female. The adults live about three years.

Gromphadorhina sp., a.k.a. "*Princisia vanwaerebeki*": This species or subspecies is only distinguishable from *G. portentosa* when major males are compared. There are at least three very different stocks labeled with this name based on the notch at the front of the pronotum above the head. This feature is highly variable by species, stock, and individual for wild and captive specimens and occurs on related genera like *Aeluropoda*. Minor males and females are physically indistinguishable from *G. portentosa*. Horns of the major male are pointed and end further back on the pronotum (reminding some of Madonna's '80s cone bra). Everything labeled "*Princisia*" has the same male genitalia as *G. portentosa*, with which they readily interbreed in captivity. Males range from 40-80mm at adulthood. Adults live three years.

Newly Born Hissers

Gromphadorhina grandidieri (tiger hisser) has the same male genitalia as *G. portentosa*. The defining characteristics are color and a wide and shallow indent at the front of the pronotum. Adult coloration is not as variable as other species, but black specimens show up now and then. Each abdominal segment is banded in black and white (males) or black and tan (females). The meso- and metathorax are black with a burgundy rectangle on each side. Adults usually live four years. They range from 40-70mm.

Gromphadorhina oblongonota (wide horn hisser): Males of this species have an elongate shape to the pronotum which extends out in front of the wide horns. Adults are a deep burgundy color. Older nymphs usually have large white rectangles on the thoracic segments. On other *Gromphadorhina,* the white rectangles occur only on small nymphs and are usually marginal. Adults usually don't live more than two years.

Elliptorhina chopardi (dwarf hisser) is a small tan and orange species that doesn't exceed 40mm. *Elliptorhina* translates as "elliptical horn." *E. chopardi*'s horns are very small and run at opposite angles along the pronotum in an ellipse or V shape. In contrast, the horns of *Gromphadorhina* species start near each side and go forward in a U shape. Adults live less than two years.

TIGER HISSER

JAVAN DWARF HISSERS

Elliptorhina javanica (Javan dwarf hisser): This very handsome and brightly colored species averages larger than *E. chopardi* and does not breed as quickly, but is otherwise similar. Adults live about three years.

Husbandry: Every container imaginable has been used to house hissers. All species climb glass and plastic. Large nymphs and adults are easy to contain, while babies are another story. Sealed containers work well but must have good ventilation. Hissers do well in dry or moist habitats, but die from high moisture with little ventilation. Though unnecessary, nearly any substrate is acceptable. Horizontal hiding areas like cork bark and egg cartons should be provided, or the roaches may spend all their time at the corners of the lid. Separate water and food dishes are used, to cut down on spoiling and pests.

The staple diet is dry kibbled pet foods, including dog and cat food. Nymphs fed high protein ferret food will be darker in color at adulthood. Low protein food can produce smaller, yellow adults. Once maturity is reached, they do not change colors, no matter what they are fed. Hissers enjoy most fruits and vegetables, as well as lettuce, rose, blackberry and other leaves.

Young reach adulthood in four to seven months. As with all animals, both genetics and nurture greatly affect adult size. Overcrowding and stagnant or overly dry conditions produce mostly tiny adults. (If they come running when watered or fed, they may be overcrowded or neglected.) Feeding only dog food or a variety of foods has only a minor effect on adult size. Though not comparable to large giant stock males, normal stocks can produce rather nice males under good conditions. Even under optimal conditions, some males will be barely an inch and a half. The tiny size suggests they're still young, but the ridge between the horns signifies adulthood. Small males will not molt again and should be removed from the culture.

A temperature range from 60-90°F is accepted, but 80-85°F is best. Admittedly, some *G. portentosa* stocks grow and breed great at 65°F. An incandescent bulb can be placed over one side of the cage to provide heat. This lets the animals regulate their own temperature and moisture. Bulbs should not be placed directly over the entire cage.

Some *G. portentosa* stocks give birth even as they're dying from poor conditions. On the other hand, other hisser species and some stocks are difficult or impossible to breed without excellent captive conditions. Gestation of the ootheca only takes two months but females can abort it, or hold onto it for years. If a culture hasn't produced babies in a year, and the hobbyist improves conditions, within a few weeks every female can give birth. Food and water should always be available. Higher ambient temperature can be the necessary factor. A warm light source for basking can more helpful. Certain foods can convince females to release their young, especially apples, pears, and oranges. Obviously a last resort, females spontaneously abort when being eaten headfirst by a predator. If the eggs are developed, she gives birth while being eaten.

Giant Cave Cockroach

Giant Cave Cockroach
Blaberus giganteus

Display: 5- to 30-gallon cage with enough ventilation to prevent mold growth. Large numbers of specimens can be kept together depending on floorspace and hiding areas, but adults will have chewed wings if food or moisture is lacking. Adults hide during the day, so a dark room with red lighting can be used, or cork bark slabs can be propped against the front of the habitat so that hiding areas are the viewing areas. Nymphs burrow, so a shallow sand substrate or no substrate will prevent them from hiding.

Background: The giant cave cockroach is one of the most spectacular of all the world's roaches, and one of the longest kept in captivity. The impressive half a foot wingspan places it squarely at the top of the list of magnificent species. Livestock was extremely difficult to acquire prior to the mid-1990s. Though it's a monster and among the most spectacular pet roaches, nymphs are easily acquired today and quite inexpensive. Adult males are notably smaller than females and are about three inches, while adult females measure in at a whopping three and a half inches. (Despite some writings, the maximum verifiable record for any roach species is just four inches.) Females above 3.5″ are exceptional but show up now and then in large cultures. Still, even a runty female has a wingspan close to six inches. (A dried specimen with the wings fully spread located alongside the live display will show off the impressive wingspan.)

The only *Blaberus* species that's nearly as large as *B. giganteus* (there is some crossover at the low end of the range) is *B. colloseus*, but the adults look different and only *B. giganteus* has the trademark orange-maroon colored nymphs.

Giant cave cockroaches are found in Central and South America and the West Indies. They happily make their homes in bat caves, but are more commonly found in smaller "caves," like hollow stumps and under logs in forested areas.

Adult males are more aggressive than other *Blaberus* as they fight, and can tear each other's legs off while sparring over a female. This behavior subsides if there are a large number of males in the cage, and not every male is so aggressive.

Husbandry: Captive *B. giganteus* do well on a diet of dry dog food and fruits. Unlike most *Blaberus,* they tend not to produce young if fruit isn't included in the diet. Apple or pear slices are quickly consumed, while some fruits like banana will be ignored if the animals aren't starving. Drinking water should always be available. With attentive care this species can reproduce in numbers, but neglect can easily wipe out a colony.

Caging should include a shallow layer of substrate for the nymphs to burrow through. 75-85°F is a good temperature range for reproduction in the captive habitat. As nymphs mature, a wide, upright, rough, vertical surface such as an upright log or sheet of cork bark should be provided. The surface can be at a slight angle but if it's far from vertical it will not be useful. This surface is very important for the molt to adulthood, or the large wings will not expand perfectly. Deformed wings not only look terrible, but greatly deformed individuals often die within a few days of the ultimate molt. Some adults die mysteriously after the final molt even if well cared for and the molt is perfect, but good husbandry reduces the percent of die-off.

DEATH'S HEAD COCKROACH
BLABERUS CRANIIFER

Display: 2- to 30-gallon cage with enough ventilation to prevent mold growth, relative to ambient humidity. Large numbers of specimens can be kept together depending on floorspace and hiding areas, but adults will have chewed wings if food or moisture is lacking. Adults hide during the day, so a dark room with red lighting can be used, or cork bark slabs can be propped against the front of the habitat so that hiding areas are the viewing areas. Nymphs burrow, so a shallow sand substrate or no substrate will allow them to be viewed.

Background: This is one of those peculiar insects whose scientific name and common name have both encountered serious confusion throughout its captive history. *Blaberus discoidalis*, a much smaller species in the same genus, was commonly traded under the scientific and common name for this species during the '80s and '90s, despite the fact it had no "death's head" markings. Someone likely knew the genus, ran across the name and thought it sounded appealing. It was much too small and looked nothing like the drawing of the real animal from the Florida Keys found in Helfer's 1964 guide.

The real death's head was something I had sought for years, but it did not seem to exist in culture. I followed lead after lead to always end up with a few more discoids and a few less dollars. I didn't come across specimens until 1998; amazingly it was a fluke, and from a major university just twenty miles from my house. I was doing unrelated work there and heard about the "cockroach room" and had a chance to check it out. I can't describe the feeling of excitement when I saw two 55-gallon trash cans full of egg cartons covered with this beautiful animal. Here they were in person, with jet black wings and a scary little face on the pronotum.

DEATH'S HEAD COCKROACH

The death's head geographic color race was originally described at *Blaberus trapezoides*. It was later synonymized by Dr. L. M. Roth with *B. craniifer*. As mentioned, this color pattern is unique to this species, however the species is not specific to the color form. (No other *Blaberus* can look like a death's head, but geographic races of *B. craniifer* can look like other *Blaberus*.)

Despite name issues, there is visibly only one death's head cockroach, and its spectacular coloration and large size make it a desirable display species.

DOMINO ROACH

DOMINO ROACH, IRREGULAR MORPH

Husbandry: Like all members of the genus *Blaberus*, it cannot cling to smooth glass or plastic, so the culture container doesn't require a lid. The habitat should include upright molting surfaces, or the adult hind wings will form incorrectly and stick out the sides. Deformed wings are unsightly, but generally do not affect survival or reproduction in this species. Dog food supplemented with the occasional fresh fruit slices is a successful captive diet.

Females bear live young which take around eight months to reach maturity. Adults rarely live more than a year. The death's head breeds fairly well and is not prone to mysterious deaths like *B. giganteus*. While rapid reproduction is possible, it doesn't breed as quickly under as wide a range of conditions as common feeders like *B. discoidalis* and *B. fusca*. Although it can reproduce in the low 70°s, 80-85°F is optimal for breeding.

DOMINO ROACH
THEREA PETIVERIANA

Display: 2- to 10-gallon cage with screen vents. Hundreds can be kept in a quart of substrate, but overcrowding can lead to sporadic severe damage during the final molt. Adults are day active but may burrow. Nymphs burrow, so a shallow, contrasting substrate or no substrate will allow them to be viewed. Oothecae should be provided substrate a few inches or more deep.

Background: This extraordinary beetle mimic from India is round and brightly marked in order to mimic a Chrysomelid beetle. Most older reference materials list this species as *Corydia petiveriana*, but *Therea* is presently considered the correct genus, though originally it was *Cassida*. Recently the available stock was relabeled *Therea barnhardti* in Germany, but *T. petiveriana* was not changed, and evidence for the new description is foggy. The spots on the cultured variety are more square than round and adults are a little larger than one wild *T. petiveriana* specimen I own, but the coloration can vary greatly and there are aberrant forms. A very similar species, also newly described, *Therea olegrandjeani*, is in culture and is similar except that it is marked with a question mark pattern. *Therea* coloration is variable and it is possible the question mark is just a geographic pattern variation of *T. petiveriana*.

Husbandry: Therea petiveriana is a member of the family Polyphagidae (collectively known as

QUESTION MARK ROACH

QUESTION MARK ROACH, IRREGULAR MORPH

desert or sand cockroaches) of which there are only a handful of species kept in captivity. Despite the usual habitat and common name, members of the family do not survive many days without available moisture. Polyphagids tend to have small appetites, and prefer rotten leaves to fruit or dog food. *Therea* adults and older nymphs also feed on banana, melon, grapes, peas, fresh sweet corn and dog or cat food. One unique characteristic of this family is that females normally bury the leathery, sculptured oothecae. In the close quarters of captivity, burrowing by adults and nymphs will bring some oothecae to the surface. Each female forms approximately 25 oothecae, each containing a dozen or so eggs. Eggs are slow to hatch and nymphs slow growing, which are common features of polyphagids. They require very little attention, but it can take years to get a culture producing, and even longer to provide continuous adults for display. *Therea petiveriana* take 3-5 months to hatch, and can take more than 18 months to reach adulthood. At high temperatures, and pristine conditions, the entire cycle, including gestation, may take as little as 12 months.

This species develops a disproportionately high number of males, which can outnumber females by as much as ten to one. Females are short-lived, and seldom continue five months after maturity. Males become adults first and rarely survive more than three months. Males have long, feathery antennae. They are much thinner and usually have a smaller diameter than females (though overall size is quite variable). Females and nymphs cannot climb up smooth surfaces, but adult males can climb almost anything.

GREEN BANANA COCKROACH
PANCHLORA NIVEA

Display: 1- to 10-gallon cage with damp substrate. Large numbers of specimens can be kept together with no issues. Adults hide during the day, so a dark room with red lighting can be used, or cork bark slabs can be propped against the front of the habitat so that hiding areas are the viewing areas. Nymphs burrow, so a shallow sand substrate or no substrate will allow them to be viewed.

Background: This small, bright green species is a handsome display animal, though it can be difficult to show off, as the nymphs are dark-colored burrowers and the bright green adults commonly hide.

This New World species is among those that have benefited from man's habit of shipping fruit, namely bananas, around the globe. *Panchlora nivea* occurs over a huge range, including Central and South America and the West Indies. It has shown up around the world and has long been established in North America. *Panchlora nivea* was probably naturalized in the United States more than a century ago. Hebard reported it was well established in Texas in 1917. This species is found along the Gulf Coast from Florida to Texas. It has been found at California ports, but for some reason doesn't seem to be established west of Texas.

GREEN BANANA COCKROACH

The green banana cockroach only survives in warmer temperate areas, but the northern range in the USA is uncertain. In colder areas here and in countries around the world, it has been encountered in greenhouses, likely transported with plants from Florida or the Caribbean. These roaches, which are strong flyers, cannot survive in homes or buildings, but may fly in from outdoor populations or reproduce in greenhouses.

Outdoors, the spectacularly cryptic, green adults can be found during the day resting between green leaves. They are capable of climbing smooth, vertical surfaces such as glass or plastic. Nymphs are dark brown and resemble the immatures of other genera like *Pycnoscelus*. However, all but the smallest nymphs can be identified because they are greenish on the underside. (Additionally, *Pycnoscelus* spp. nymphs can climb glass.) The 1st instar young of this live-bearing species are tiny at 1.5mm, approximately the size of a fruit fly. The immatures are incapable of climbing smooth, vertical surfaces, and prefer to live a subterranean life.

A unique physical character of this species is possession of specialized barbs on the rear legs of the adults. All cockroaches have spines on the legs, but *P. nivea*'s can break off and cause sharp, rapid pain. It's a rare occurrence, but if you handle adults regularly, occasionally a spine will pierce your skin and cause pain that feels like a bee sting but only lasts two seconds (not the feeling of a pin prick as with other species).

As with most of the popular pet roaches, *Panchlora* are blaberids, so are live-bearing. Two to three dozen young huddle around the mother until they have changed from white to brown. They like to nibble on rotten wood, leaves, fruit and dry dog food. Damp substrate should be provided for burrowing nymphs, while flat pieces of bark can be offered for adult hiding spots.

PANCHLORA SP. "GIANT"

Display: 1- to 10-gallon cage with damp substrate. Large numbers of specimens can be kept together with no issues. Adults hide during the day, so a dark room with red lighting can be used, or cork bark slabs can be propped against the front of the habitat so that hiding areas are the viewing areas. Nymphs burrow, so a shallow sand substrate or no substrate will allow them to be viewed.

Background: Whenever there's a very handsome-looking but tiny cockroach like the green banana cockroach (*Panchlora nivea*), there's a tendency to think, "Wow, that would be a really great species if it weren't so small." As if in answer, before the turn of the century, a much larger species of *Panchlora* showed up in culture. It's more than twice the size of *P. nivea* and is a notably

Giant Green Banana Cockroaches

lighter shade of green. There doesn't appear to be any information on where the first stock originated, beyond the tropical New World. There are about fifty *Panchlora* species found across the tropical Americas and West Indies and without an exact collection location it may be impossible to find the correct species name. (Specimens have been sent to taxonomists with no luck so far.) Like *P. nivea*, males are noticeably smaller than females, but the size discrepancy is far more exaggerated in *Panchlora* sp. "giant." Females (at 27mm) are many times the mass of males (18mm); measurements being taken with vernier calipers from the front of the pronotum to the end of the tegmina.

Husbandry: Adults are excellent flyers and climbers, while the brown nymphs are earthbound and prefer to hide in the upper half inch of substrate. A damp area in the substrate is important for the health of nymphs. Extremely tiny young, the size of small fruit flies, hatch after ova have developed inside the mother's body for a few months. *Panchlora* species tend to fare best if a partly rotten log is placed in the cage. They eat the rotten wood but their primary diet consists of dried dog food or an equivalent, and most fruits.

Glowspot Cockroach
Lucihormetica subcincta

Display: 1- to 10-gallon cage with damp substrate. Large numbers of specimens can be kept together with no issues. Adults hide during the day, so a dark room with red lighting can be used, or cork bark slabs can be propped against the front of the habitat so that hiding areas are the viewing areas. All stages burrow, so a shallow sand substrate or no substrate will allow them to be viewed. Rotate specimens on and off display.

Background: This species started out as *Brachycola subcincta* and was later revised as a member of an earlier described genus, *Hormetica* (1838). In 1999, *Hormetica subcincta* along with another eight members of the genus *Hormetica* (fifteen species presently remain in *Hormetica*) were moved, along with the description of a new species in the genus, *Lucihormetica fenestrata*. In 2003, *Lucihormetica grossei* was described, bringing the number of members to ten (eleven as of Nov. 2008, I. Fritzche).

The species of *Hormetica* that were moved to the genus *Lucihormetica* all have one characteristic in common: males have a pair of luminescent organs on the front of the pronotum. "Luci" means "light." These are the only luminescent orthopteroids known (grasshoppers, cockroaches/termites, mantids, etc.). Unfortunately, dead specimens and captive-reared individuals lack this ability. Whether there's a rare chemical or luminescent microorganism missing from the captive diet is unknown. However, in captivity some diets can change the color of the spots between yellow and orange after maturity. (Dietary changes affecting coloration during internal construction of the exoskeleton is not rare in cockroaches, but after the fact is unique.)

Males and females are sexually dimorphic. Male coloration is shiny and bright. At maturity, the pronotums are not only colored differently but are also structurally dimorphic. The female has a smooth, rounded pronotum, while the male's has upraised ridges (similar to male hissing cockroaches). As usual, the abdominal segmentation is dimorphic, but this is only noticeable in the later instars. Other sexual characteristics aren't visible prior to the ultimate molt.

Lucihormetica occur primarily in the northern half of South America, *L. subcincta* being native to Columbia. It is found in tropical forests, commonly not too far from streams or other sources of fresh water.

Husbandry: This species is capable of surviving dry periods, but lack of available moisture leads to damaging molts and death. A section of the substrate should be kept wet at all times. Natural temperatures normally range from 75-88°F and but rarely go as low as the mid-sixties. The wild diet is unknown, but in captivity, the usual fares of dry dog food and fruits are accepted. As with many species, apple is a favorite, and reproduction is rare without fruit. The natural life history is not documented, but it's likely similar to captivity, up to two years to reach maturity, with adults living one to four years. Although the majority don't live so long, multiple four-year-old females have been known to produce young. In other types of cockroaches, females that live overly long far outlive their reproductive viability.

Ornate Velvet Roach
Deropeltis paulinoi

Display: 2- to 20-gallon cage with bark for egg laying. Large numbers of specimens can be kept together with no issues. Adults and nymphs often hide, but can be encouraged to sit in the open by not offering dark hiding spots.

Background: The genus *Deropeltis* contains a number of large, egg-laying species found primarily in Africa. Males are slight with long wings, while females are physically similar to large nymphs. Nymphs are commonly black with orange markings or legs, while adults tend to be solid black

Glowspot Cockroaches

ORNATE VELVET ROACH, SUBADULT PAIR

ORNATE VELVET ROACH, MALE

or brown. *Deropeltis paulinoi* is a medium-size species found in Angola, Botswana, Congo, Namibia and Zambia. Adult females of this species retain the coloration of late instar nymphs, which are predominantly black with orange markings. (Early instars have a reverse pattern, mostly orange with a large dark spot.) Males are jet black like other species. Adults are 32-38mm and the lighter-bodied males tend to be the longer of the two genders, due to the wings.

Life Cycle: Females produce dozens of large oothecae, which are often glued under objects. They should be misted occasionally and kept at 75°F or higher, to hatch after three to four months. Although oothecae do not hatch with total neglect, they are far sturdier than other *Deropeltis,* and hardier than most egg-laying species.

Deropeltis paulinoi is an extremely productive species, but it probably wouldn't be eaten by the majority of insectivorous vertebrates. The orange and black nymphs have warning colors for a reason. *Deropeltis* emit quinones as a defensive secretion. The secretion is similar to quinones produced by giant round millipedes. The pungent chemical smell is noticeable, but the quantity of defensive fluid is minimal compared to giant millipedes. Although nymphs are a useful food for waterscorpions, mantids and tarantulas, it's important to use forceps during feeding. Forceps are clumsy and difficult, but they are needed to prevent fingertips from being brown for the next few days.

Unique and superb coloration, along with appreciable size, ought to make this a highly sought species. However, reproductive ease and ready acceptance of overcrowded conditions ought to make it an excellent feeder insect. The problem is that quinone production greatly detracts from its value as a feeder. Nevertheless, the ornate velvet cockroach is one of the sharpest-looking species available for displays.

ORIN'S COCKROACH
DORYLAEA ORINI

Display: 2- to 20-gallon cage with bark for egg laying. Large numbers of specimens can be kept together with no issues. Adults and nymphs often hide, but can be encouraged to sit in the open by not offering dark hiding spots.

Background: This beautiful species was described in 2002 by Dr. Luis M. Roth of Harvard University (Blattodea: Blattidae, type number MCZT_35085, MCZ type database @ Harvard Entomology), but was in culture a few years before it had a name. Originally it was labeled as an unknown species of *Periplaneta* from Malaysia. It can be difficult or impossible to find scientific names for unidentified roaches. It's extremely rare for a species not to have a name, but it is common to not be able to locate the name.

Unless it's a well-publicized species, the only method is to send male and female specimens to a taxonomist. The male specimen is most important since its dissected genitalia are a key indicator. Identification isn't the easiest task since there aren't very many taxonomists, and each specializes in specific groups (as often determined by geography as systematics). Even if a taxonomist can be found for the type of roach to be identified, the roach may not be identifiable without exact collection location information. It may not be an easily identifiable species even using dissected genitalia, or the information may not be readily available. Most important of all, the taxonomist must be willing to offer his or her time.

ORIN'S COCKROACH, ADULT

Simply running across an unknown species doesn't mean much. I've sent many cockroach specimens to various taxonomists that probably will never be looked at (only after asking if it's OK to send beforehand, and for their preference on how specimens are preserved). I've sent milipede specimens that had been previously cataloged but not described; even if they were someday described, it would have nothing to do with the specimens I sent. With *Dorylaea orini*, I was fortunate in many ways. My primary interest was to find the correct name, but to have "discovered" a new cockroach was both a surprise and a thrill.

ORIN'S COCKROACH, IMMATURE

Dr. Roth graciously offered to take a look at some dead specimens for identification. He determined this one was a *Dorylaea* species and checked for previous descriptions. Later, he said he would formally describe this new species and kindly asked what name I would like. I suggested *pulchra* (meaning "beauty" since the nymphs are quite striking) but he said there were dozens of roaches with the species name *pulchra* so it wouldn't be a useful or good name. Then he asked if I wanted to use my last or first name for the species. (I chose my first name, but may have decided differently if my name were John, Mike or Bob.)

Dorylaea orini is a member of the cockroach family Blattidae. The genus *Dorylaea* contains a few dozen species from tropical Asia. The nymphs start out brown, but become strikingly marked in black and white banding within a few molts. Nymphs are highly attractive and rival any other species. Adults have long, dark-brown wings, and are not among the more attractive roaches. They grow to 26-34mm. Nymphs and adults can climb glass, but are easily kept in check by a petroleum jelly barrier.

Husbandry: Females produce dark brown oothecae containing approximately thirty eggs. They are glued lengthwise to rotten logs, and tiny bits of log are chewed up and carefully glued over the exposed side. Oothecae will be dropped, left exposed on the substrate, if there are no good spots for laying. This species does not cannibalize its own oothecae, which makes high captive-reproduction rates easy.

This cockroach hails from the tropical country of Malaysia but it doesn't require extremely high temperatures. It does well at room temperature around 72°F, but grows more rapidly at higher temperatures. Some dryness is acceptable, though moisture should be available for drinking at all times and eggcases need to be misted regularly unless the humidity is above 75%. Good ventilation is required and microscreen should be used for vent covers to help maintain humidity if ambient humidity is low. The animals do well on a diet of dog food and will also feed on many fruits and vegetables.

Remarks: This species is easily maintained and visually appealing. It sits out in the open under some conditions. It is a good feeder for certain animals but is kept almost entirely for observation. *Dorylaea orini* usually borrows its trade name from *Eurycotis decipiens*, but I kind of prefer the name Orin's cockroach.

ZEBRA COCKROACH
EURYCOTIS DECIPIENS

Display: 2- to 20-gallon cage with bark for egg laying. Large numbers of specimens can be kept together with no issues. Adults often sit out in the open while nymphs usually hide.

Background: Eurycotis decipiens is one of the older captive species, but remains among the more spectacular display animals. The adult's dazzling coloration and outgoing behavior are major reasons for its favor. It can prove a challenging species to keep over multiple generations, but the stock seems to have grown accustomed to captive hardships over the last few decades. The cultured stock was originally collected in Costa Rica.

ZEBRA COCKROACH, ADULT

Zebra roaches' bright coloration and bold behavior result from its defensive abilities. They produce a defensive secretion like their commonly kept relative *Eurycotis floridana*, with a primary active ingredient of gluconic acid. However, their defense isn't as potent and they are a great deal less likely to use it. Like *E. floridana*, the adults are easily differentiated from older nymphs by the presence of short, square front wings (there are no hind wings). One or both wings are commonly shed if the animal is handled roughly, similar to a lizard shedding its tail.

Life Cycle: Females produce large bean-shaped oothecae that are abandoned soon after hardening. The cases are usually dropped on the cage floor, but only because there isn't an adequate piece of wood provided to which the female would normally glue cases. Egg cartons aren't adequate. These hatch after a few months (temperature dependant). Another trait shared with *E. floridana* is that the 1st instars are able to cross over a vertical surface of petroleum jelly unless the barrier is more than a few inches wide and well maintained. Hatchlings are large and easily contained by normal window screen. After the first molt, glass climbing is easily eliminated with petroleum jelly. Teflon paint is not a reliable barrier for this species. Nymphs hatch out black, but the white and black banding (as the common name *zebra* suggests) begins to develop after the first molt. Older nymphs can be striped or black, but adult coloration is not variable.

Growth to adulthood takes four to six months and adults can live up to a year. Dog food is useful as a staple diet at all stages of growth. Fruits and vegetables are also eaten. Cultures fare best in the 75-80°F range. Temperatures below 70°F shorten adult lifespan and can result in oothecae that do not hatch. High temperatures can increase growth but tend to create desiccation problems, especially for the oothecae.

Pink Cockroach

Pink Cockroach
Gyna capucina

Display: 2- to 5-gallon cage with little substrate or hiding spots. Adults are best shown in small displays and rotated. Nymphs stay buried and die without substrate.

Background: This elegant cockroach hails from tropical Africa. Adults range in size from 27 to 38mm and are the largest *Gyna* species kept in culture. The diamond-shaped pronotum extends partly over the abdomen and is painted a dark maroon with some lighter markings. The underside of the adult is a peculiar orange-pink color. The outer wings (tegmina) appear pink, but upon very close inspection it is possible to see extremely thin alternating bands of red and white. The adult's antennae are banded in orange and black. The nymphs are a chocolate brown with twenty or so large yellowish-tan spots.

Two other *Gyna* are commonly cultured. *Gyna bisannulata* looks similar except it is much smaller, the adult has a white margin at the front of the pronotum and the wings are a patchy purple-pink. It is easier to keep than *G. capucina*. *Gyna lurida* nymphs are black and white, looking very different from the other two species' nymphs, but are similar in shape and habits. *Gyna lurida* adults are usually porcelain white with brown markings, but cultures can be chosen for yellow. This species is extremely productive in captivity.

Husbandry: It is a member of the family Blaberidae, and like all members of this taxon, is ovoviviparous. That is, it gives live birth after having held eggs internally during development. This genus holds the record for large numbers of eggs contained in one ootheca, with up to a few hundred born at one time. As a result, the eggs are tiny and first instar immatures are extremely small compared to the adults. Despite the large number of young it is capable of having, it is difficult to culture. Large broods are uncommon and the survival rate can be very low. Females often produce no young. The only way I've been able to encourage production and survival is by offering the females fresh peach slices. The shiny, torpedo-shaped nymphs are very active and become adults in six to nine months at 75-85°F.

Pacific Beetle Mimic Cockroach
Diploptera punctata

Display: 2- to 10-gallon cage with no hiding areas. Large numbers of specimens can be kept together, though only the adults exhibit mimicry.

Background: Despite a muted and unassuming appearance, *D. punctata* is one of the most fascinating and unique cockroaches in the world. Physical and olfactory mimicry and reproductive strategy set apart this handsome species. It's a native of Asia but has been spread around the tropics by man. In the U.S., it is widespread in Hawaii. *Diploptera* is the only genus in its subfamily. There are six other members of the genus *Diploptera* but none of them have been kept in the hobby.

Visual mimicry is heightened by olfactory imitation. As the common name suggests, the adults look very much like a beetle—a medium-size, brown tenebrionid (darkling beetle). The wings are formed in such a way as to suggest a straight line down the middle where a beetle's elytra meet, rather than the overlapping tegmina of the common cockroach. This cockroach is capable of discharging a defensive secretion which smells (and probably tastes) the same as the common darkling beetle defensive secretion. It does not smell the same as quinone secretions of spirobolid millipedes or *Deropeltis* cockroaches.

Diploptera punctata's defensive secretion is of similar chemical composition but is minimally potent compared to some darklings. The secretion is ejected from the second pair of abdominal spiracles and is created by glands surrounding the trachea. It is composed of *p*-benzoquinone and two of its derivatives. Darklings produce a mix of benzo- ethyl-, and methyl quinones from glands near the end of the abdomen.

Wyttenbach and Eisner suggest that *Diploptera punctata* is unusual among roaches because the female mates immediately after a molt, before her exoskeleton hardens. Her secretory glands don't function for the first few hours and they suggest the male offers some protection, although he does not discharge the defensive secretion during mating if she is attacked, only when he is himself disturbed.

It's an interesting concept, but the majority of roach keepers are aware that this is a common behavior for most, if not all, of the blaberid species kept in captivity, many of which have relatively impotent defensive smells, or lack them altogether. The behavior is likely a result of mate competition, having nothing to do with defensive glands.

What sets this species apart from most invertebrates, making it a singular species among the cockroaches, is its reproductive methodology. It is a member of the family Blaberidae, or live-bearing roaches, but just as a few live-bearing roaches are not in this family, *D. punctata* is not live-bearing. It is the only known viviparous cockroach. That is, the female nourishes underdeveloped eggs internally until they are ready to be born. The large relative size and low number of nymphs per brood are byproducts of this strategy.

Viviparity: "Only one cockroach, *Diploptera punctata* (Eschscholtz) is viviparous. The female produces a very small ootheca which is rotated and retracted into the uterus. Initially the eggs do not have sufficient yolk and water to complete development; the embryos obtain solid nourishment and water from the mother. There are only about a dozen eggs in an ootheca and the embryos compete for nourishment in the uterus; if the number of eggs are reduced experimentally, e.g., by removing one ovary prior to oviposition, the size of the smaller number of new-born nymphs are much larger than the greater number of control nymphs." (Dr. L. M. Roth, excerpted from *Invert-Magazine*. Dec. 2002.)

Husbandry: The Pacific beetle mimic is easy to keep in captivity. It reproduces when kept as low as 70°F but is more productive at 75-85°F. Despite small broods, it can reproduce quickly enough to be used as a feeder. Moisture should be provided in the form of a water dish or damp areas of the substrate. General fare of dog food is eaten but fruit can enhance reproduction just as with most of the live-bearing cockroaches.

Pacific Beetle Mimic Cockroach

Lobster Cockroach

LOBSTER COCKROACH
NAUPHOETA CINEREA

Display: 1- to 10-gallon cage with damp substrate. Large numbers of specimens can be kept together with no issues. Adults hide during the day, so a dark room with red lighting can be used, or cork bark slabs can be propped against the front of the habitat so that hiding areas are the viewing areas. Nymphs burrow, so a shallow sand substrate or no substrate will allow them to be viewed.

Background: This species has been commonly kept as a feeder since the late 1990s. It originated in Africa, but the stock kept in the United States was collected from the 'wild' in Florida. As a result of human activity before good entomological records, the lobster cockroach has long been found throughout the world in tropical areas and even greenhouses in cold areas. Though widespread, it is not a commonly encountered species so it is possible all stock kept in the United States is descended from the original Florida source. This species is listed in comprehensive pest control identification guides as an occasional pest of greenhouses, and is reportedly said to feed on *Panchlora nivea* found in the same greenhouses. The occasional occurrence in greenhouses is likely, but the predation report is highly dubious.

While some people believe the name "lobster" results from its epicurean taste to reptiles, the name comes from the markings on the pronotum that someone thought looked like a lobster. (This takes some imagination.) The name "cinerea" is from the Latin word *ciner,* meaning "ashes" and most certainly refers to the gray adult coloration. The nymphs are brown. This is the only species in the genus *Nauphoeta.*

Husbandry: The live-bearing *N. cinerea* female produces about two dozen young after forty days of gestation. The male looks similar but is slightly smaller, and can be sexed even as a small nymph by the small posterior ventral abdominal segment. Both genders have wings about as long as the abdomen. Reproduction and growth are affected by temperature, food, and cage size. (Overcrowding can result in cannibalism of new molted individuals.)

All stages can climb glass, so they are generally offered to predators in small cages or from feeding dishes, with the upper sides coated with petroleum jelly to prevent escape. Teflon paint also works well with this species. Newborn nymphs are many times larger than a fruit fly but are useful food for any tarantula spiderling. Adults are rarely more than twice the mass of a feeder cricket, which limits food use for large reptiles.

RHINOCEROS COCKROACH, ADULT MALE

RHINOCEROS COCKROACH
MACROPANESTHIA RHINOCEROS

Display: 5- to 20-gallon cage with shallow, damp substrate and vented lid.

Background: This animal is widely known as the heaviest cockroach in the world and is also the most expensive. Display specimens can be a few hundred dollars, with pairs going for even more. The reason the price will likely remain high is that they grow slowly, reproduce minimally, are often killed by simple easily-avoided mistakes, and are rarely imported. Even if they were regularly imported, they command a decent price in their home country of Australia. However, the expense is not without benefit, as they live a long time. These massive specimens can live seven years after reaching adulthood and the time to maturity can take up to six years.

Husbandry: Full-grown specimens are easy to keep healthy on a diet of decaying hardwood leaves, fresh fruit and dry dog food. Nymphs eat the same, but must be provided a shallow substrate for molting. In nature they build extensive tunnels terminating in an open cave which is difficult to duplicate in the type, density and volume of substrate provided in captivity. (A structured display could be built without substrate, as is often used for ant and mole rat displays.) Other

RHINOCEROS COCKROACH, FEMALE WITH 1ST AND 2ND INSTAR YOUNG

RHINOCEROS COCKROACH, MOLTING

burrowing roaches burrow rather than build dens, and can molt successfully with substrate on them. Rhinoceros cockroaches suffer massive deformities and often die when buried in dirt during a molt. They can survive a bit of dryness, but moisture in the form of damp substrate should always be available. They are not built to climb and will not be able to get into water dishes to drink nor past heavy cage decorations. They can be kept from 65-82°F without trouble, but temperatures above 85°F are reported to kill them in short order. Short periods down to 40°F don't seem to cause lasting damage.

This is a live-bearing species and females usually produce approximately two-dozen young. It takes a year or two following the final molt before females are capable of having young, and at least a year in between. Males should be removed if young are expected, since they cannibalize the young. (In nature, the females kick them out of the den, but the male is forced to keep coming back in a display.) Females can be kept with their own nymphs for up to six months, but should be immediately removed if any young disappear. Immatures do not cannibalize even if they are a number of molts apart. It will take them three and a half to five and half years to reach maturity, with larger specimens and females requiring the longest times. The final molt takes the longest time, up to eighteen months.

BEETLES

EASTERN HERCULES BEETLE
DYNASTES TITYUS

Display: 1- to 10-gallon cage with damp substrate and very limited ventilation. Coconut fiber is an excellent substrate as it retains needed moisture and lets the adults exoskeleton stay shiny and clean. It should be kept less than 1" deep to keep specimens in view. Adults fight over food, but are usually kept together, so the extremely rare death can occur if there is a very large male. The monstrous larvae can also be a memorable display.

Background: The eastern Hercules beetle, *Dynastes tityus,* is found in the United States, from New Jersey, south to Florida, and west to Texas. It can be locally common but is difficult to find in most areas. The larvae require giant, old, dead, compost-filled trees, that have been removed from most areas. Adults are active between the months of June and September and fly to bright blue-white lights just after dusk.

The relatively plain female possesses only tiny bumps on her head, while the flashy male is armed with a huge horn on both the head and dorsal surface of the thorax (middle body segment). There is a tiny horn on each side of the thoracic horn, which may serve to protect the eyes. The two large horns form a working pincher used in duels over food and females. Up and down movements of the head control the pincher. Although the horns are formidable, fights seldom result in more than superficial damage (scratches and dents). Only the very largest of males are capable of inflicting mortal damage.

Dynastes tityus is one of the most colorful rhinoceros beetles. Adults range from army green to yellow by individual and stock. The elytra are covered in variable maroon to brownish-black

EASTERN HERCULES BEETLE

EASTERN HERCULES BEETLE

spots. They are capable of rapid physiological color change resulting from exoskeletal structure. The entire body turns nearly black when wet and returns to normal when dry. Depending on ambient conditions and age of the specimen (older exoskeletons react more quickly), the change can take as little as sixty seconds.

Life Cycle: Like most creatures, the eastern Hercules beetle begins life as an egg. Each ovum is white to yellowish in color and is a little over an eighth of an inch in diameter. The outer shell of the egg is thin, permeable to water and prone to drying out. For this reason, it is important to keep the egg in moist substrate.

The adult female is very particular about where she deposits her eggs. She tends only to target tree holes with certain qualities. Tree holes are the hollow centers of very old (often dead or mostly dead) trees that are filled with years of accumulated and composted dead leaves, rotten wood and forest litter. Many of the best tree holes were once home to raccoons or owls that added discarded food and frass to the compost. Unsheltered tree holes are inadequate because of widely varying moisture levels that would either dry out or drown the eggs and larvae. Small tree holes are also not acceptable because they cannot provide enough food for the gigantic grubs. Furthermore, the consistency of the compost inside a tree hole is very important; it must form easily into clumps.

In captivity, a tree hole habitat is recreated by placing a mixture of compost manure, crushed rotten wood, and pulverized dead hardwood leaves (1:1:1 ratio) into a five or ten gallon aquarium with a glass or plastic lid. The make-up and consistency of the substrate is not only important for egg laying, but is also important because the substrate is the grubs' food. A glass or plastic lid is seldom completely airtight and a minuscule gap provides plenty of airflow. Additional holes or vents cause the substrate to dry out.

The female deposits a single egg into a large clump of substrate, which she has spent hours compacting with her legs and body. The female has a retractable ovipositor only seen during egg laying. As the egg is laid, the ovipositor creates a cavity twice the diameter of the egg. The extra space is needed because the ovum absorbs moisture and expands as the larva inside develops. Next, the female pushes and packs more dirt over top. She is now finished with her motherly care and goes on to the task of preparing substrate for the next egg.

Eastern Hercules Beetle, Male Pupa

The fertilized egg expands and is full-size in just a few weeks. Interestingly, an infertile ovum appears the same at first, but distinguishes itself as it almost never expands noticeably. Near the end of the third week, the spiracles and tips of the larva's jaws can be seen through the transparent shell. Another few days and the egg hatches.

The grub is an eating machine. It rapidly consumes the surrounding substrate and grows to hundreds of times its mass in a few short months. To begin, a tiny white larva (about the size of a bb) hatches from the egg after three weeks. This small grub begins to feed on the surrounding substrate and quadruples its size within a week. Two more weeks and the young larva goes through its first molt, and the grub's head, which had been only a little over a sixteenth of an inch in diameter, becomes a quarter of an inch across. This second instar larva grows quickly to become the size of a pinkie finger. After six more weeks, the grub molts again and the new head is half an

EASTERN HERCULES BEETLE, EARLY 3RD INSTAR

inch across. This third instar larva develops into a monster white grub bigger than a golf ball! Many spectators find the frighteningly monstrous grub every bit as fascinating as the beautiful adult.

The larva requires little care but should be dug up and checked on every month. The first inspection should take place three to four days after hatching. Most difficulties and deaths are seen—and should be watched out for—during the first instar. The body of a healthy larva is mostly white with a bluish-black or brownish abdomen. If grubs do not appear plump and grow noticeably within the first week after hatching, there is something wrong with the substrate and the larvae must be placed in new substrate. During the monthly check-up, if the food appears to be mostly pellets or powder, it should be changed so the grub does not starve. Another reason for a monthly probe is the substrate surface may appear normal but be dangerously dried or soggy within. Lastly, a piece of dry dog food can be buried in the substrate (for extra nutrition) after each check-up.

The time between the last molt and formation of the pupal cell is highly variable. Poor substrate, low temperatures, genetics and overcrowding contribute to lengthening this period. The time required between larval molts is consistent unless conditions are poor. If quality substrate (food) is not provided, larvae can take years to develop. Long periods of low temperatures (40°F -55°F) do not damage the larvae but will slow or stop growth. Larvae from different genetic stocks—less commonly the same stock—often pupate at a different time under identical conditions. Full-grown grubs should not be kept together at a rate of more than two grubs per gallon of substrate or they may have too little food, damage each other, and disturb each other during pupal cell formation. Depending on the above factors, it may take only four additional months to as long as two more years for the larva to form a pupal cell following the last larval molt.

The enormous, full-grown larva gradually changes from white to yellow before pupating. When ready, the grub uses its body and jaws to construct a smooth-walled chamber deep in the dirt. This cell is about the size and shape of a chicken egg. Next, inside the pupal cell, the larva begins to shrink and the legs and head become immobile. Finally, after three to four weeks, the grub molts to become a pupa, which looks like a tan, mummified adult. The incredible monster pupa cannot move its legs or head and is only able to wiggle its abdomen.

In a few weeks, the legs, thorax and head of the pupa begin to darken. Another week and the outer husk collapses and out comes a full-grown beetle. At first, the pronotum is black and the outer wings are white. The elytra turn brown and

Eastern Hercules Beetles, Mating Pair

then, after a week or two, the armor changes to the army green to yellow adult coloration. The young adult is vulnerable for the first few weeks because the carapace takes time to harden. Dropping a new adult or placing it with mature adults frequently leads to abdominal damage and death. Once the shell has hardened, the beetle is exceptionally difficult to harm.

Most rhinoceros beetles, including wild-caught *D. tityus*, only live two to five months as adults. A unique aspect of the eastern Hercules beetle is that captive-reared adults live six to twenty-three months, as it is the only member of its genus to hibernate. Diapause lasts two to three months at 75°F, but can be extended to eighteen months with no ill effects at 52°F. After diapause is over they continue for three to five months. The adults can still be handled and are able to walk around while in diapause, but they do not eat, lay eggs or fly. After hibernation is over, the beetles begin to mate, lay eggs and feed on beetle jelly or brown sugar water.

Mating is accomplished by placing a male and female in the same cage for a few days. Mates locate each other easily, even in large cages. In captivity, females spend most of their time underground. The male's horns get in the way of digging, so he has limited ability to chase her under the substrate. Unmated females surface each night in search of a male, while mated females remain in the substrate. (Mated females may spend much time on the surface if the substrate is too shallow or inadequate for egg laying).

Fighting Rhino
Xylotrupes gideon

Display: 1- to 10-gallon cage with damp substrate and very limited ventilation. Coconut fiber is an excellent substrate to keep adults clean. It should be kept less than 1″ deep to keep specimens in view. Adults fight over food, though are usually kept together, so superficial damage can occur.

Background: The mighty Kwang (pronounced *kwon*) is the quintessential fighting beetle. Ounce for ounce, professional beetle trainers consider it the strongest of all the world's fighting beetles. If you've seen one of the TV shows with footage of beetle fighting, you've seen mighty Kwang.

Adult males are outfitted with large, opposable horns designed for fighting over mates, territory and food. Only males with the largest horns are used for competitions. The majority of

FIGHTING RHINO

wild males are left alone since only the largest win battles. During tournaments, the males are usually fought on a hollow log containing a female so that they have something to battle for. The beetle trainer taps-waves a small stick to control the beetle. The stick doesn't touch the beetle but somehow guides the small fighter. Fights are divided into rounds, are governed by a number of set rules and are judged in a manner similar to boxing. (Most of the rules are identical to Thai boxing.) Beetle fighting is a truly humane sport since the heavily armored, thick-shelled males cannot harm each other, and only toss their rival out of the way.

Thailand is the world capital of beetle fighting. Structured tournaments are a part of Thai heritage. In northern Thailand a huge festival and parade are arranged around the annual tournament. Included in the festivities is a pageant where young women compete for the title of "Miss Kwang." In the mid-1980s beetle fighting was temporarily banned because of illegal gambling. (It may be hard to believe, but fortunes were won and lost on a beetle fight.) The ban was short-lived and the Thai government presently encourages celebration of this ancient tradition.

Adult Kwang live four to six months and must be fed to ensure a normal lifespan. Beetle trainers in Thailand leash their charges (when not training or fighting) to a big chunk of sugar cane so the beetle can drink from the sap. They eat most beetle jelly flavors and varieties, but shy away from jellies with added protein. (Protein jellies are used primarily for female stag beetles.) They will also drink sugar water. Bananas and most fruits should only be used as a secondary food or avoided altogether, or the beetles won't live as long.

The Mighty Kwang's Relatives: The scientific name for this beetle is *Xylotrupes gideon.* It is a member of the family Scarabaeidae, tribe Dynastinae. The common name "princess beetle" originated in Japan and dates back to the late 1960s. *Xylotrupes* is found across a huge expanse that includes Australia, Borneo, Philippines, Taiwan, Indonesia, New Guinea, India, China, Sulawesi and Malaysia.

Two other members of the genus look very similar but have slightly different finishes to the exoskeleton. The species *Xylotrupes pubescens* has long been listed as a separate species and is easy to differentiate only because the beetle's upper surface is "hairy." Major males lack a cephalic denticle (small horn or process on the middle, upper surface of the head horn). *X. pubescens* lives alongside *X. gideon* in the Philippines and Sumatra. *X. florensis* is found only in the Lesser

FIGHTING RHINO PAIR

Sunda and Tanimbar islands of Indonesia, including Flores Is., Alor Is. and Timor Is. It is very similar to *X. gideon*, but beyond genitalia diagnosis, the dorsal surface is more glossy and females have a shiny pronotum. Major males also do not have a cephalic denticle.

Xylotrupes gideon has been considered the most wide-ranging rhinoceros beetle, with an expansive range extending through India, China and across the Indo-pacific islands all the way to Australia. Due to an expansive range and great variation in horn structure between and across populations, there have been more than 30 names awarded to *Xylotrupes*. In general most have been found to be subspecies or synonyms. Recent phylogenetic analysis suggests that five species (each extending across only a part of the '*gideon*' range) make up the complex. These are *X. florensis*, *X. gideon*, *X. meridionalis*, *X. mniszechi* and *X. ulysses*. (Accordingly, the Thai species would be *X. mniszechi* and the commonly reared stock from Borneo would be *X. gideon*.) The research is based on detailed measurements of the anatomical features of dried specimens, primarily the male genitalia and population horn allometry. The new taxonomic structure has yet to be investigated through crossbreeding experimentation. It would be significant if any or all "species" result in viable crosses and if crosses retain fertility in further generations. It is a very easily cultured beetle—and speedy for Dynastinae—which would make results readily obtainable, but the acquisition of verified livestock and time involvement requirements may perpetually stall more conclusive evidence.

The maximum adult male length from the front of the horn to the end of the elytra is about 2.5" for most species/subspecies of *Xylotrupes*. However, the smallest reach a maximum of 1.5" and the largest grow to more than 3.2".

Farming the Mighty Kwang: Cultivating *X. gideon* is easier than for other rhinoceros beetles—easier even than *Dynastes tityus*. One female lays 40 to 100 eggs in almost any wet mat material, even potting soil or dirt. As usual, females delay laying till there is a preferable medium. Unlike other Dynastinae that hold many or all eggs till death, *Xylotrupes* end up depositing eggs even in substandard medium when death approaches. Eggs have a good hatch rate and larvae survive poor and minimal feeding that other Dynastinae couldn't live through—though, underfeeding can still cause death and results in substandard adults. Rotten wood and decayed leaves are premium foods, but grubs even grow on moldy pine needles. A limited amount of dog or cat food is accepted to supplement nutrition provided by quality substrate.

Larvae molt twice and grow to become huge grubs—about the size of *Dynastes tityus*, but much more hairy. The substrate should be damp and moldable so they can build chambers where they will safely change to beetles. In only six to ten months the larvae construct pupal cells and three weeks later become pupae. Adults emerge from the pupae after about three weeks.

The bulkiest males are the most likely to have long horns, but rarely small males have long horns. Little males with large horns seem to result from a shift to increased food quality during third instar.

TRICERATOPS BEETLE
PHILEURUS TRUNCATUS

Display: 1- to 2-gallon cage with damp substrate and very limited ventilation. Coconut fiber is an excellent substrate to keep adults clean. It should be kept less than 1" deep to keep specimens in view. One animal per cage. Adults should only be housed together for breeding.

Background: The triceratops beetle, *Phileurus truncatus*, is a large, peculiar, carnivorous scarab. Its habitat extends throughout much of North America: parts of Canada, Mexico and most of the U.S. It is a member of the family Scarabaeidae (dung beetles, chafers, rutelines, etc.), subfamily Dynastinae (the rhinoceros beetles).

TRICERATOPS BEETLE, MALE

Determining the gender of the larvae does not appear to be possible, though determining the gender of adults is easy. Most rhinoceros beetle larvae can be sexed using the presence or lack of tiny marking visible towards the rear of the abdomen. Though a less reliable method, the males or females of many species are differently-sized and this difference can be readily seen in the size of full-grown larvae. Larval gender of *P. truncatus* grubs is indeterminate because the males and females are identical in size and there are no notable sexually dimorphic light or dark markings visible on the rear ventral segments. Adults, on the other hand, are easy to sex, though at first glance it doesn't seem that way since both genders have horns. The male's terminal abdominal sternite is rounded while the female's is pointed. Both genders have the same size horns, but the female's are pointed at the tip while the male's terminate in a small globular structure.

Husbandry: Grubs grow rapidly, taking just a few months to reach full size. They are gregarious, and a dozen can be reared safely in a 1-gallon container. Pupal cells are usually all made alongside each other. When reared together, any stragglers will pupate at the same time and can emerge much smaller than siblings. It is certain that in the wild any larvae which do not pupate along with others in the same tree hole, face instant death as a food source for freshly emerged adults. Other rhinoceros beetle larvae such as *Dynastes tityus* can be induced to pupate with hormones of larvae in the same container, but not prematurely.

Larvae feed on finely ground, rotten, hardwood leaves mixed with a small amount of rotten wood and compost. Supplemental foods should never be used, as they result in larger size accompanied by wing deformations that ensure rapid adult cannibalism and subsequent loss of the culture.

TRICERATOPS BEETLE LARVA

In nature, the adult female is often found with small larvae and may protect them by eating competition. Adults make squeaking stridulation noises and vibrate, while larvae vibrate strongly but silently through rapid pulsations of the thorax. This allows young and parent to communicate and they may stay together through the late instars in nature. In captivity they should be separated by early 2nd instar to prevent cannibalism.

Adults live 1.5-2 years after emergence, at room temperature with decent feeding. They may live much longer if dormant cool periods are provided and counted as part of the span. Adults nibble on some sweets, but are predators—they will become very aggressive towards tankmates and die prematurely if offered only fruit or sugary liquids. They prefer large, slow-moving prey, especially large beetles, but may catch crickets, mealworms and a variety of insects depending on the cage. *Zophobas* pupae are an easily provided food source.

TRICERATOPS BEETLE PUPA

HARLEQUIN FLOWER BEETLE
GYMNETIS CASEYI

Display: 1- to 10-gallon cage with humid substrate and limited ventilation. Numerous adults can be kept together without aggression. Coconut fiber is an excellent substrate to keep adults clean. The beetles are day active, but substrate should be kept shallow to keep specimens in view.

Background: Most keepers are familiar with the commonly cultured *Pachnoda* and *Eudicella* flower beetles from Africa. Few are familiar with any of the hundreds of tiny U.S.-native flower beetles (mostly around 1/4"). However, we do have a few impressive species. The harlequin flower beetle, *Gymnetis caseyi* (formerly known as *G. flavomarginata*), from Texas and Louisiana is the third largest flower beetle in the U.S., slightly smaller than *Cotinis texana* and *Osmoderma eremicola*. The velvety black body marked with bright yellow "flames" combine to make this arguably one of the most beautiful of all the flower beetles.

Like the eastern Hercules beetle, this species is designed for physiological color change, and the yellow flames change to greenish black when damp. The exoskeletons of older beetles react more quickly. Displays should not be excessively damp, or the pattern and beauty of the beetle is obscured.

Husbandry: The beetles eat banana slices and cut grapes. Beetle jellies can also be used. These items should be closely monitored and greatly reduced if pests are observed on the food. Adults live three to eight months and are extremely productive egg layers once they start. Females will refuse to lay eggs if the substrate qualities are not adequate. It is helpful to save a volume of larval frass to mix in the substrate to induce egg laying. If larvae or adults are not separated, the large numbers of burrowing larvae will stop adults from laying more eggs.

HARLEQUIN FLOWER BEETLES

In the wild, larvae are often encountered in compost heaps full of plant matter. Nutrient-rich heaps may hold mite-covered, sickly grubs. The larvae feed primarily on decomposing leaves, but can also be fed a smaller portion of rotten wood, compost, or other foods. Leaves and wood should be finely ground, almost to a powder, for best results. Mixed hardwood leaves are the only required food. In addition to being finely ground, the leaves should have been on the ground for at least a few months. Supplements such as dog food pellets can improve size and growth. If dog food is fed instead of carefully prepared substrate, or if more is offered than can be eaten in a day or two, it will only cause health problems and death.

The time from egg to adult is five to twelve months, depending on food quality and temperature. They are temperate and can survive cold, but the habitat should be kept at 72°F or higher unless slow growth is desired. Low temperatures stop growth and make it easy to stagger a colony in order to insure adults year round. The first generation should produce enough larvae to split into as many groups as desired. Food quality greatly affects the speed of growth, but is far more difficult to regulate since leaves from different areas vary according to what trees live there.

HERMIT BEETLE (OR THE HERMIT FLOWER SCARAB)

Display: 1- to 10-gallon cage with humid substrate and limited ventilation. Numerous adults can be kept together with no aggression. Shredded, damp, light-colored wood is a good substrate. The beetles can be active day or night, so substrate should be kept shallow to keep specimens in view.

Background: Osmoderma eremicola is the largest flower scarab species found in the United States. This distinction is a direct result of the small size of beetle fauna in the U.S., as only the largest southwestern *Cotinis* species grows nearly as big (maximum 30mm). Adult female Hermit Beetles are known to grow to 32mm. Wild adults are usually in the 25-32mm range, but larger specimens may exist. Captive-produced adults can be as small as 15mm and still be reproductively viable. Size is influenced by genetics, but key determining environmental factors include food, moisture and space.

Osmoderma eremicola occurs across much of the eastern, central and western United States and Canada. Specimens found west of the Mississippi River are sometimes labeled a distinct species or subspecies, *O. e. subplanata,* but are identical to eastern *O. eremicola. Osmoderma scabra* (18-25mm) is a very similar but smaller species that occurs in much of the same range, east of the Mississippi River. I've collected both species in different ends of the same rotten log. *O. scabra,* in areas I've collected, is far less common. Its wing-cases aren't smooth like *O. eremicola,* but are covered in wrinkles and have a purple sheen. *O. eremita* is a well-known, common European species that appears very similar to *O. eremicola,* but the sculpturing on the male's pronotum is different. (The male's head sculpturing is the same.) There are eleven additional members of the genus: *O. brevipenne, O. caelestis, O. coriarium, O. cristinae, O. dallieri O. davidis, O. italicum, O. lassallei, O. opicum, O. richteri* and *O. sikhotense.*

HERMIT BEETLE, MALE

Osmoderma eremita has a huge range across Europe, from France to Russia, and includes thirty countries. It is considered rare and is protected in some overdeveloped European countries. However, except where every trace of habitat has been replaced by manmade structures and agriculture, it can be very common. The usual trap methods rarely work, which makes this very common beetle seem rare. Beetles are almost never collected by ferment, light or pitfall traps, and must be directly collected. *O. eremicola* can likewise seem very rare but direct manual collection reveals they are ridiculously common.

The common name *hermit beetle* is an old and established name, but it's hard to know why this name was chosen. A hermit is a recluse, but members of this genus tend to be colonial and gregarious in nature in the grub and beetle stages. As mentioned, even where *Osmoderma* species are extremely common, the adults are rarely encountered outside their rotten wood habitat, which may be the reason for the reclusive name. The same common name is used for *O. eremita* (*eremita* and *eremicola* both mean hermit) and *O. scabra.* For differentiation, *O. eremicola* is uncommonly listed as hermit flower beetle, but of course this name applies to all the hermit beetles since they are in the flower beetle subfamily.

Life Cycle and Husbandry: The genders are notably distinct as the pronotum, pygidium and head are shaped differently. The head and pronotum differences are much more pronounced on large males. The male's pronotum is larger in proportion to the body and deeply sculptured. The male's head has a large ridge along the dorsal surface of each side, which is not present on females. The pygidium (the last dorsal abdominal segment) is triangular and pointed backwards on the female, and is visible from above. The male's pygidium is rounded and not notable from above, independent of the size of the male. As usual for Cetoniinae, the male grub has a tiny, central, black dot visible through the ventral surface of the abdomen, towards the end. Genders develop, emerge and live similar lengths of time. Due to the pygidium shape, females are longer than males of the same size.

Adults need to feed to survive more than a few weeks, but eat little compared to some Cetoniinae. Since they rarely leave their rotten log, it's hard to say what they feed on in nature. Hungry beetles will chew on moistened kibbled dog food, but do not cannibalize small larvae. They will feed on cut fruits such as banana, but do not consume much, and these foods can lead to scavenger mite outbreaks unless very careful attention is paid to removal of leftovers. Only maple syrup/sugar water or commercially prepared beetle jellies are recommended, as they rarely lead to mite problems.

Grubs can feed on the same food as most flower beetles, including rotten hardwood leaves, various compost and rotten hardwood. However, they have a strong leaning towards rotten wood. If fed on primarily rotten leaves like other flower beetles, they'll grow slowly, become stunted, or die. Hardwood that yields the best results is old, very soft, damp and brown (not extremely dark brown or black). 75% rotten hardwood and 25% compost is recommended.

Captive habitat requirements are minimal. Adults can be kept in a cage as small as 16oz (80% full of substrate), but a few gallons is preferred. The egg-laying substrate should be 25-40% compost, with the rest being ground rotten wood. Rotten wood should be ground up as finely as possible, as larvae cannot eat hard rotten wood. Grubs can have a difficult time eating soft rotten wood if it is offered as large chunks or logs. Substrate should be very damp, but not waterlogged. Ventilation should be minimal—around 2% of total lid area—to limit drying out. No stage fares well under dry conditions. Larvae rarely damage each other, but more than a

Hermit Beetle 2nd and 3rd Instars

dozen 3rd instar per gallon is not recommended. Remove grubs in late 1st to 2nd instar so females can continue to lay eggs.

In nature, in the north, the normal time required from egg to beetle is approximately two years. White, 2mm eggs are laid in summer, hatch in three to four weeks and molt to 2nd instar after another month. The 2nd instars grow over the next month or so and then form hibernation cells to spend the winter as the temperatures cool. In mid-summer, the 2nd instars molt to 3rd instar which form pupal cells in the fall. Adults emerge around April and lay eggs in summer. Decent colonies will contain eggs, 1st, 2nd and 3rd instars and beetles in mid-summer (primarily second instars and pupae in early spring).

Captive rearing can take much less time than in the wild. Temperature has everything to do with the life cycle. If the temperature is 70°F or lower, it will take about two years for the cycle. Even at 70°F, late 2nd instars normally form overwintering cells and hibernate. At 75-80°F, depending on food quality and other factors such as population density and moisture, the time to maturity is 8-12 months.

Pupation in this species is similar to other Cetoniinae, in that larvae form earthen cocoons. These special cells can be removed from the substrate (unlike the cells of other scarabs) but must never be damaged. Grubs cannot form a second cell, and survival rate without a perfect original cell is extremely low even if the correct type of fake cell is provided. Pupation time in this species is peculiar in that it can take as little as one month, to more than six. Mature 3rd instars commonly form pupal cells at the same time, forming them right next to other cells. At 75°F, total time in the cell is usually three months, but one or two often come out early and emergence is staggered.

Adults live three to five months, and are active. They don't play dead like Gymnetini. Adult beetles are capable of flight and are sometimes collected at lights in the wild, but flight doesn't seem to ever occur in captivity.

This is one of the few large native beetles that are easy to keep in captivity indefinitely and can be easily produced in large numbers if desired. It can take some experience to perfect rearing methodology and even more to produce especially large specimens. Important rearing points: rotten wood is a necessity, moisture level should be very high (but not so high the substrate goes anaerobic), and damage to pupal cells must be avoided.

Elephant Stag Beetle
Lucanus elaphus

Display: 1- to 10-gallon cage with damp substrate and very limited ventilation. Coconut fiber is an excellent substrate to keep adults clean. It should be kept less than 1" deep to keep specimens in view. Males generally do not hurt females, but males can damage or kill each other and should be kept apart.

Background: Lucanus elaphus is a gaudy monstrosity with no rivals in all of North America; only the comparably small-jawed *Pseudolucanus* species are even in the ballpark. Beyond the obvious visual appeal, beetle collectors prize the elephant stag beetle for another reason: it is difficult to get. Larvae are easy to rear, but it is rare in culture because it is difficult to keep through successive generations.

Husbandry: It is possible to breed *L. elaphus* in reasonable numbers with knowledge of the parameters of each stage in the life cycle. The primary difficulties lay in convincing females to deposit eggs, and in timing captive-reared adults to insure fertility. The most subjective and complex aspect is providing a proper environment for egg laying and hatching. A few inches of compost is placed in a 10-gallon aquarium, rotten logs are placed on top and then mostly covered in finely ground rotten wood. Rotten oak logs are suggested, but many other hardwoods have been

Elephant Stag Beetle, Male

used successfully. No matter how carefully chosen, a seemingly perfect log may not work. Of course even if conditions are perfect for laying, tons of infertile eggs are valueless. Adults live nearly a year following the molt to maturity, but the active mating period is seldom more than a few months. The female lives months longer than the male, so it is possible for him to die or become too old and feeble before she is ready to emerge from hibernation. Neglecting to provide a period of lowered temperature (50°-60°F is

Elephant Stag Beetle, Pupa

good) will normally throw them out of sync and can also shorten their lives to half a year.

Elephant stag beetle larvae are surprisingly forgiving, as they can survive poor substrate that would kill most other stag species. Supplemental feeding is important to the production of gigantic males, but if wood is mixed with a large amount of compost, leaves or other material, the larvae tend to die or result in very small or unhealthy adults. To improve adult size and survival rate, all pests in any media used should be destroyed through heating. Rotten oak, beech, willow, and locust have been used successfully to feed larvae, but other hardwoods are likely acceptable.

Males fed correctly during the larval stage are measurably and impressively bigger than poorly fed males. Females are approximately the same size independent of feeding. Females are small, normal-looking beetles less than 1 1/4″ long, while the large jawed males can grow to more than 2 1/2″. Despite the incredible size difference of the genders, both male and female larvae, in fact, molt only twice, which is the rule for all stags (Lucanidae).

ELEPHANT STAG BEETLE, PAIR

Eggs and pupae should not be dug up or handled. If accidentally disturbed, the pupa stage is undemanding and requires only a slightly concave surface and 100% humidity. The beetle lives about a year after freeing itself from the pupal skin, while the ovum, larva, and pupa stages combined normally take a year.

WESTERN BLACK STAG
PSEUDOLUCANUS MAZAMA

Display: 2- to 5-gallon cage with damp substrate and very limited ventilation. Coconut fiber is an excellent substrate to keep adults clean. It should be kept less than 1″ deep to keep specimens in view. Adults do not usually harm each other when kept at about one per gallon, but might fight under some conditions.

Background: This lucanid is the third largest North American stag beetle (North America boasts the overall smallest Lucanidae of any continent other than Antarctica, which has none). It is found in Arizona, Colorado, New Mexico, Utah, and in northern Sonora Mexico. Wild adult males don't exceed 35-36mm, but a captive-bred male (from captive-reared parents) grew to 39mm (out of only ten males total, F1 and F2). With such a small number of males reared, it's hard to estimate the maximum size a male might reach in captivity. Females grow as big as 33-34mm.

In the wild, larva have been found in decaying Arizona sycamore, *Platanus wrightii,* and in decaying ponderosa pine, *Pinus ponderosa*. Adults can be collected on the trunks of the sycamore trees at dusk, and will come to mercury vapor and black light.

There are two other North American members of the genus *Pseudolucanus*: *P. capreolus* and *P. placidus*. *Pseudolucanus capreolus* grows to 42mm. It is common and widespread, except in the west, and is especially common along the Appalachian mountain chain and foothills. Getting *P. capreolus* to lay eggs is very challenging, but rearing larvae is straightforward. *Pseudolucanus placidus* reaches 35mm and is relatively unknown in captivity due to lack of livestock.

Egg Laying: Pseudolucanus mazama is by far the easiest North American stag beetle to culture, mostly because females lay eggs readily. Wild and captive females willingly lay eggs in common lucanid setups: well-rotted but firm beech and oak logs buried in damp, finely-ground rotten wood or layered mat. Given the natural food range for grubs, females will lay eggs in any type of well-decayed wood. Acceptance of a relatively wide range of humidity and level of decay is unusual for a stag. A 10-gallon aquarium with solid glass lid is acceptable for two or three females. More females will result in the same number or fewer eggs, though it's hard to determine if they eat each other's eggs, accidentally smash them or simply interrupt preparation. One to three males can also be placed in the cage to insure fertility. In a 10-gallon, there is no notable aggression between males or against females.

It is best not to disturb eggs since handling can kill them. Three months after the adults have died is a reasonable time to wait before digging through the breeding terrarium (if offspring have not already been observed through the glass).

Grubs: Grubs are sturdy and seldom die from handling. (Larvae of the tiny North American stags like *Ceruchus* and *Sinodendron* die within hours if touched, though the large grubs of *P. capreolus* and *L. elaphus* are almost as sturdy as *P. mazama*). Larvae look similar to those of *L. elaphus*, but the head capsule is more orange and the V-shaped molt line has a notably wider angle. Annoyed grubs stridulate loudly and can be heard from twenty feet away. They rub a specialized comb on the trochanter of both rear legs against a similar comb located on the tibia of the middle legs. Stridulation doesn't often occur during handling, rather shortly after the grub is returned to its cage or when the cage is moved.

Larvae feed primarily on soft, light brown, yellow or white rotten wood (dark-colored wood is generally too far decayed). Substrate should always be thoroughly damp, but not soggy. Close attention should be paid to sterilization of substrate components. Tiny substrate pests can obliterate the food value, as well as stress and kill grubs. Nematodes and mites can be difficult to see, but when present in number massive or complete die-off near pupation is common. Larvae often look fine and healthy up to that point.

Grubs can be reared in groups only if there is adequate space and food. Cannibalism is minimal if properly set up, but separate containers will ensure the highest survival. The minimum size cage for a single larva is eight ounces. Larvae should be moved to roomy containers or separated before 3rd instar to prevent cannibalism from taking a high toll.

Adults: Normally, adults emerge in summer, but stay in the pupal cell until the following year. They can be handled and appear normal, but they

MALE STAG BEETLES
(TOP) *PSEUDOLUCANUS MAZAMA*
(MIDDLE) *PSEUDOLUCANUS PLACIDUS*
(BOTTOM) *PSEUDOLUCANUS CAPREOLUS*

do not, eat, mate, or dig (except to rebury themselves). The adults live 12 months, but only actively mate and feed during the last four. It might be possible to knock them out of hibernation early through excessive disturbances.

Beetles will feed on brown sugar water, but have a minimal appetite. The wild diet of adult *P. mazama* is uncertain and even the females don't seem to be actively interested in feeding on other insects. Many tropical stags are voracious feeders (females especially) and kill large, slow-moving insects to lap up the body fluids. It's possible adult *P. mazama* compensate for a lack of correct feeder invertebrates in their natural setting by subsisting primarily on reserves built up during the larval stage, however lack of voracity may be an artifact of the diapause required for the subtropical life cycle.

Adults can be kept together without incident in a ten-gallon with a few inches of substrate and some logs. Nevertheless, many other stag species kill each other in close quarters. It is best to keep only one adult if they have to be placed in a tiny temporary cage such as a film capsule or small deli cup.

Male or Female: Gender determination for adults is straightforward since females have thick, serrated jaws, while even minor males have sickle-shaped jaws. The male's jaws are smooth and have a single inner tooth. Third instar larvae are easy to sex with certainty. The female's ovaries are visible through the sides of the abdomen and resemble small, tan beanbags. The ovaries are not attached to the outer wall and move with the grub so it may take a few moments to locate them.

Life Cycle: Eggs are laid in late summer, hatch after three weeks and larvae reach 3rd instar by December. Pupal cells are formed the following late spring. Like all Lucanidae, the grubs molt twice (3 instars) and the next molt results in a pupa. Grubs should be moved from the egg-laying cage to rearing containers when the first 2nd instars are noticed. The growth period, including egg, larva, pupa and molt to adult takes approximately twelve months.

WESTERN STAG BEETLE LARVA

Poor quality food or pest problems in the substrate can slow down development and change normal time periods. There is a small window of time for the 3rd instar to develop into the pupa. Although internal timing isn't perfect and diminishes with each captive generation, the beetle is aware of the calendar despite consistent photoperiod and temperature in captivity. If food was substandard and it is too late in the summer, the mature grub may still build a pupal cell but will remain a larva until the following summer. Despite the fact no feeding takes place for nearly a year, this period doesn't seem to affect the adult beetle's health or size—as long as the grub was healthy to begin with. The beetle then goes through the normal hibernation bringing non-feeding time to more than twenty months. *Lucanus elaphus* does the same thing. The majority of scarabs and stags delay development if they miss their season, but accompaniment with yearlong lack of feeding in the larval stage is very peculiar.

Lastly, temperatures above 72°F should be avoided after the pupal cell is formed, but before pupation. High temperatures can cause them to flee their cells.

DARKLINGS

Display: 2- to 30-gallon cage with damp substrate and good ventilation. Coconut fiber or sand substrate is preferred, as soils result in dirty specimens and buildup on feet. Adults do not harm each other and can be kept in high densities with no problems. Glass or clear plastic tubes can be used to provide hiding spots.

BLUE DEATH-FEIGNING BEETLE LARVA

Background: Every insect keeper is familiar with the common yellow mealworm, *Tenebrio molitor*, used to feed reptiles, assassin bugs and other small predators. However, few people are acquainted with its many relatives.

The darkling family (Tenebrionidae) is one of the largest groups of beetles in the world, with about 1500 species described in North America alone. There is the yellow mealworm, the barely visible, grain-infesting, confused flour beetle (*Tribolium confusum*), red flour beetle (*T. castaneum*), the large feeder mundoworm (*Zophobas morio*), and a variety of large desert species.

There are various differently shaped and sized beetles available in the United States. What darklings lack in color, they make up for in outlandish body structure. Many are easily reared following the same techniques outlined below for *Eleodes*. There are large, exquisitely shaped and alien-looking pie dish darkling beetles found near Roswell, New Mexico. There are also a handful of rotund, large-headed *Asbolus* species including the blueberry-colored and bumpy *Asbolus verrucosus* found in the southwestern U.S.

Eleodes are personable invertebrates that do not bite, pinch or sting. They are hardy, active display subjects. The natural defenses include an extremely hard carapace and a mildly offensive acidic fluid squirted from the end of the abdomen. The defensive fluid is relatively harmless to humans and is meant to disgust would-be predators with a pungent smell or, as a last resort, a horrible, sour taste. Some *Eleodes* species almost never utilize the defensive fluid and only if they are never handled and therefore easily frightened.

Husbandry and Life Cycle: Adult darkling beetles are long-lived and easy to feed in captivity. *Eleodes* beetles regularly live two to three years, but can survive longer. A record adult male *E. suturalis* survived for twelve years in captivity, although by that time all the feet had fallen off. Beetles enjoy feeding on cut fresh fruits of all kinds and tear ravenously into pieces of dry dog or cat food. The beetles are not predatory, but opportunistic and eat most things including a variety of dead animal parts.

The cage is the first concern. A lidless bucket, an aquarium, or plastic cage with a vented lid works fine. A lid is not necessary for desert species, since they cannot climb glass and do not possess flying wings. The outer wings are thick, incapable of flight and cannot even be opened. If a lid is used it must be well vented or the adults will die prematurely. A cage for egg-laying should have at least two inches of potting soil, compost, or rotten wood on the bottom (gravel, sand, or no substrate is fine if larvae aren't desired). A section of the substrate should always be kept moist. The adults do not require additional moisture, although it doesn't hurt to mist them from time to time or offer water in a shallow dish. Remember they cannot climb smooth surfaces and will drown in a deep dish that is not filled with gravel or sponge.

Eleodes females burrow into the substrate to bury their eggs, often a few inches down. One female can easily produce over a thousand eggs during her long life. The oblong eggs are softer and more delicate than silkmoth eggs, but are harder and more durable than most beetle eggs. The tiny, white, cylindrical eggs of various *Eleodes* species are around a millimeter in length. Eggs hatch after a few weeks into minute yellow grubs.

Eleodes larvae appear nearly identical to common yellow mealworm larvae. Of course, the full-grown mealworms are much larger than ordinary mealworms. The grubs are the same light yellowish brown color and possess similar shaped heads and legs.

DARKLING BEETLE, *ELEODES SUTURALIS*

Most darkling grubs cannot be kept alive in a container of dry rolled oats like common yellow mealworms. *Eleodes* mealworms are no exception. The larvae must be kept in slightly moist substrate or they will slowly dry up and die. The occasional apple or potato slice just doesn't provide enough moisture. Moist potting soil, peat, and compost are all acceptable mediums. Mix these 50/50 with aspen bedding to improve larval development. *Eleodes* mealworms feed on many things, including oatmeal, dog food, cracked corn, and leaf mulch buried in the substrate. The food tends to mold excessively if not consumed after a few weeks and should be replaced accordingly.

At last, the mature larvae stop feeding and carve themselves a rather large pupal cell out of the surrounding dirt. This cell is designed to protect them during the drastic transformations to adulthood. The period between the pre-pupa stage and adult stage is the most difficult time for the beetles and without adequate space for pupal cell formation, many will die or become deformed. Only during this vulnerable period do they require some distance from their tank mates. It is important to provide each larva with a minimum of two square inches of surface area as well as two inches of depth. After a few weeks inside the cell, the mealworms become immobile and curl into the characteristic pre-pupa C shape. Then, the larvae pupate. The whitish tan mummy-like pupae aren't much bigger than superworm or giant mealworm pupae, but the legs, head and pronotum of each pupa are relatively gigantic.

First the eyes and mouth parts, then the legs and pronotum of the pupa gradually transform from light tan to nearly black. Next, the gigantic adult drags itself out of the old skin. The rather large elytra (hardened outer wings) of the beetle stretch out and form a huge dome over the scrawny abdomen. For the first few hours the elytra are milky white in color but still mostly see-through. At this point it is easy to see how hollow the new adult really is. The space around the small abdomen, large head and legs explain why the mediocre larvae become such gigantic adults.

The young adults stay inside the pupal cell for a week or two as the soft elytra and abdomen harden into the sturdy adult carapace. As soon as they dig their way up to the surface, the adults can be placed in a cage for egg laying. Mating is

ELEODES SPINIPES PUPA

simple. The male hops onto the female's back and often hangs around long afterwards. The male and female adults are easy to tell apart because the male is comparatively thin with parallel sides, while the female has a more bulbous body shape.

Females may deposit eggs immediately, but usually wait a few months after emergence before producing ova. If the females do not lay eggs after three or four months, two or three half-grown yellow mealworm larvae can be placed in the substrate. Hormones produced by the growing mealworms induce egg-laying in these related species.

Pie Dish Beetles

Blue Death-Feigning Beetle

Pie Dish Beetle
Embaphion muricatum
These are extremely productive and can be easy to rear. Adults usually live a few years with good care. The beetles eat a variety of items, but relish dry pelletized food formulated for juvenile bearded dragons. Too much humidity or extreme dryness will cut their lives short. Beetles have no defensive fluid nor offensive odor.

Superworm Beetle

Other Darklings

Blue Death-Feigning Beetle
Asbolus verrucosus
Although data is lacking, these appear to live many more years than *Eleodes*. Captive breeding has been limited primarily due to sparse larva production. Larvae have a paddle-shaped terminal segment and are paler than *Eleodes*. Beetles have no defensive fluid nor odor.

Superworm or Mundoworm Beetle
Zophobas morio
Larvae are available at most pet shops or through mail order. Place each mature larva in a cup and it will transform into a pupa within a few weeks.

ELEODES SPINIPES (MALE) AND *ZOOPHOBAS MORIO*

The beetles are over an inch long and make a nice display when other beetles aren't available. Beetles possess very potent and readily released defensive fluids.

EYED CLICK BEETLE
ALAUS OCULATUS

Display: 2- to 5-gallon cage with damp substrate and limited ventilation. Coconut fiber is an excellent substrate to keep adults clean. It should be kept less than 1″ deep to keep specimens in view. Adults do not harm each other and it's difficult to overcrowd them.

Background: Long ago, I was patiently tearing apart a log with the back of a hammer, looking for bugs, when I saw a strange, unidentified larva. It looked somewhat like a giant ant lion. It had a black head, sickle-shaped jaws and a plump, tapered, straw-colored abdomen. I went to pick it up and its jaws clamped onto the skin next to my thumbnail. It couldn't have deliberately chosen a better place to inflict pain. For what seemed like one of the longest moments of my life it repeatedly squeezed, loosened ever so slightly and then squeezed harder. It took every bit of my will power not to smash it with the hammer. I had never seen anything like this large-jawed giant grub and there was no way I was going to lose it.

I've never held my hand against a hot stove for as long as I could, but I'm sure it would be similar. After a very long time my arm refused orders from my brain and flew back, sending the grub flying. I searched the area with no luck. The log was also thoroughly searched for others. For years I wondered what that strange grub was. Surprisingly, it turned out to be the larval stage of what had always been one of my favorite beetles.

Six or seven years later, I found some stumps containing two of these mystery grubs. I was thrilled to see them again. I put them in separate deli cups, and they chewed their way out that night. I moved them to super-thick, hard-plastic screw-top containers—and they chewed their way out. Their mandibles are incredibly more powerful than those of other beetle larvae. The grubs are only contained successfully by glass or metal. A month or two later, they settled down and one successfully pupated and changed to an eyed click beetle. (The other died during one of its escape attempts.)

EYED CLICK BEETLE

I was surprised, especially since the eyed click beetle was always in my top ten list of beetles. It's the biggest beetle I've ever found in my own yard, but it's not easy to find here. I've only found half a dozen adults over decades, and generally one at a time. I've always wanted to try

EYED CLICK BEETLE PUPA

to breed this beetle, but scarcity of livestock and lack of husbandry information has prevented it.

Alaus are members of the family Elateridae, also known as click beetles because of a mechanism they use to right themselves when they get stuck on their backs. *Alaus oculatus* also uses this to startle predators—and keepers, too—when grabbed. There is a second common name for this family, the name for the larvae: wireworms. The larvae of most species earn the name because they look like a piece of wire (or a streamlined version of a mealworm). I have reared a number of native click beetles, including a shiny black species that reaches 30mm, and even the largest had streamlined, typical, wiry larvae. Being familiar with wireworms, I hadn't considered the mystery grub one of them.

Life Cycle: I found half a dozen grubs a few years later that were medium and large. All six pupated within a month and it was apparent the smaller grubs were not immatures, just smaller individuals. (Females average larger, but one of the small ones was a female.) I put the adults in a 2.5-gallon aquarium with a few inches of damp rotten wood flakes mixed with compost and some wood chunks. The adults ate sugar water and lived between four and five months. I saw some tiny, 1/16″ long, cylindrical eggs but never saw a hatchling. I found just one small, 1/4″ grub a month after the last adult died. It was around twenty times the size of the eggs, which is still quite tiny. I threw it in my mixed darkling/isopod/cricket culture and figured it could chew on the wood and dog food. I didn't really expect to see it again.

Months later, I lifted up a log in the cage to find a half-grown *Alaus* grub chewing on half an *Eleodes* larvae. I later saw it eating crickets, isopods and many more *Eleodes.* (At this point I realized its siblings might not have simply failed to hatch.) Its appetite was huge. It was puzzling that a grub found deep inside rotten wood in nature would have entirely predatory habits in captivity. All logs they were found in were somewhat dry and pretty much devoid of other insect life. It wasn't till this larvae was full-grown that I learned the answer. At maturity it chewed a hole straight into a solid log in the cage and settled down to pupate. I remembered then I had never actually found an immature grub inside a log.

Remarks: Eyed click beetles are handsome, large, somewhat long-lived and active, which makes them a great display animal. The requirements for the various stages listed here along with livestock and time investment might allow for successful culturing. Still, it might not be possible to rear *A. oculatus* in decent numbers without an inordinate amount of space and time.

BUTTERFLIES AND MOTHS

MONARCH BUTTERFLY
DANAUS PLEXIPPUS

Display: Caterpillars are great for displays, but adults do not do well in cages. 2- to 30-gallon cage with no substrate and very good ventilation.

Background: No matter what part of the world a person lives in, they are probably familiar with the monarch butterfly, *Danaus plexippus*. This incredible butterfly is found on every continent and most islands where milkweed grows. The monarch flies great distances across land and sea and colonized most of the earth by its own power eons ago. Few, if any, other single arthropod species could compare. Only over the last few hundred years, with man's ability to travel globally, have some arthropod hitchhikers been able to enlarge their range as expansively as the monarch.

MONARCH CATERPILLARS

The monarch butterfly is not only famous for its ubiquitous range but also for its migration. In North America, adults fly as far as two thousand miles each fall to spend the winter in a few cool mountains in Mexico. In early fall, groups of a few dozen or more can be seen resting on branches

MONARCH BUTTERFLY

across the U.S. and Canada as they make their way down to their winter stomping ground. The large groups are an amazing site, but cannot compare to the final stop where every tree seems to be covered in leaves made of butterfly wings.

Adults are large butterflies with a wingspan of four to five inches. They are magnificently colored in deep black and orange, and appear nearly identical wherever they are found. Remote island populations are relatively isolated, with the lone straggler from the mainland only showing up every few hundred or thousand years. Roughly ten percent of the monarchs found on the Hawaiian Islands are white, where they should be orange.

The two-inch caterpillar is as magnificently colored as the adult. The body is circled in alternating bands of yellow, black and white, and both the head and rear have a pair of long, black tubercles resembling antennae.

I remember watching monarchs kept as a class project in elementary school. I can't recall the grade or teacher, but vividly remember the look of the strange caterpillars and the bright-green, barrel-shaped chrysalis transforming to black and orange while the butterfly inside developed.

MONARCH CHRYSALISES

Life Cycle: The eggs are incredibly tiny (about a 30th of an inch) and as green as the leaves, but are not too difficult to locate because they are elongated and project off the surface. Wild females carefully attach one egg to one leaf, no more than two on any plant. If you encounter an egg, take it inside, because its predators are legion. The chance of finding it as a caterpillar a few days later is close to zero.

I've reared dozens of different native sphinx moths and giant silk moths each summer and am always impressed by the caterpillars' rapid speed of growth compared to most other invertebrates. However, the monarch's speed of growth was a surprise. They make the moths seem slow. It takes only two weeks for the egg to hatch, develop through all five larval instars and transform into a chrysalis. Another eight days or so and the first adult butterfly emerges.

Rapid growth means the caterpillars are only available for display for a short period. One positive aspect of rapid development is that caterpillars eat far fewer leaves than a similar-sized moth caterpillar would consume.

Monarch butterflies are much larger than the usual butterflies and skippers in the area, and it is striking how huge they look up close. The wingspan of a decent adult is right around five inches. They are bigger than most of the exotic butterflies found in butterfly exhibits and about the same size as many of the native giant silk moths.

One time I attempted to keep a few adults. The best cage I could find was made of cloth screen and measured 14 x 14 x 24 inches. I assumed that, like moths, they would quickly beat their wings into worthless tatters if I put them in a glass or plastic cage. I stood the cage on end and hung a hummingbird feeder inside. The feeder was filled with "hummingbird food" because it is formulated to approximate flower nectar. I hoped they would be willing to feed from the fake flowers. The cage was a bit on the small side, but they seemed content to flutter from one side to the other all day long.

After a few days I began to worry. I hadn't seen them drink from the feeder. I also worried they might become restless and beat their wings off in the small cage. They never damaged their wings, but after ten days, the first adult died. Since I couldn't get them to eat and they were from my area, I decided to let the remaining three go. Maybe one or two would make it to Mexico in the fall.

PAINTED LADY BUTTERFLY
VANESSA CARDUI (LINNAEUS)

Display: Butterflies should be given at least a 5-gallon net cage. Caterpillars can be kept in small containers but require good ventilation.

PAINTED LADY BUTTERFLIES
FEEDING ON BEETLE JELLY

Background: Adults of this handsome, moderate-size butterfly range from 2-2.5" in wingspan. The painted lady is a member of the lepidopteran family Nymphalidae, which includes various angelwings, mourning-cloaks and the viceroy. It's a cosmopolitan species and is often said to be the most widespread butterfly species in the world. This species has likely become the most widespread caterpillar found in classrooms and kept in homes. Painted lady butterfly kits are always available through biological supply houses, but are also seen for sale at garden shops, science stores, educator supply shops, the local grocery store and even at a close-out stores. The kit includes a net cage for the adults and a card to send away for caterpillars. Half a dozen or so caterpillars arrive in a cup with enough artificial diet to finish their life cycle.

Foodplant: Canadian thistle is an invasive plant found in flower beds and along roadsides across North America. It's likely anyone reading this is familiar with the look of this plant, if not the name. The plant is painful to pull from the ground without thick plastic gloves because the tiny needles break off under the skin and have to be dug out later. Even when more than a foot of the root is pulled, it still grows back. It even grows back after being treated with all but the strongest herbicides. It's nice to know something eats this terrible pest plant. The caterpillars in the photos were fed strictly on Canadian thistle. Caterpillars are also supposed to feed on other thistles, sunflower and hollyhock.

Ova: Ova are tiny little blue specks about the size of the period at the end of this sentence. It's important to keep the leaves fresh or the eggs

PAINTED LADY BUTTERFLIES
(TOP) EGGS AND HATCHLINGS
(MIDDLE) CATERPILLAR
(BOTTOM) CHRYSALIS

quickly dry out and die. They change from blue to black as the caterpillars develop inside. Hatching occurs just 48 hours after egg are laid.

Caterpillars: Like the eggs, young caterpillars prior to fourth instar can die rapidly when the leaves dry out. Foodplant cuttings should be kept in a dram vial or other plastic container filled with water. A hole is poked in the lid of the vial and the stem wrapped with small pieces of paper towel to prevent curious little caterpillars from drowning. Cut thistle only lasts about four days before it starts to turn yellow and dry out. A fresh cutting should be placed against the old piece when it first begins to look a little yellow or wither. The caterpillars will crawl over to the fresh leaves. The larger caterpillars are most likely to find the fresh leaves. 1st or 2nd instar may need to be carefully placed on fresh foodplant.

There are four molts, five instars. 1.5mm black 1st instar caterpillars begin to chew small holes in the underside of the leaf. The top layer of the leaf is left untouched, so it looks like the leaves are being converted into wax paper. (They only eat the bottom layer until the last instar.) The very hungry caterpillars eat constantly and molt every 48 hours. 2nd and 3rd instar caterpillars spin silk webbing around the feeding area, and will drop and hang from a silk thread when disturbed. 4th and 5th instar caterpillars drop to the ground and curl up to become a spiny ball. 5th instar is obtained in approximately eight days. 5th instar is an extremely hungry caterpillar, and can be difficult to keep up with. After three days the caterpillar climbs to the roof of the cage or under the base of a foodplant leaf and spins a silk pad. The back legs are hooked into the pad and then the caterpillar hangs head down until it molts into a chrysalis.

Chrysalis: The chrysalis is tan with metallic gold spots. It wriggles energetically when disturbed, but can't go anywhere. This stage requires almost as much time as all the previous stages combined, about eight days. A few days before eclosion, the coloration of the adult butterfly begins to be visible through the transparent shell of the chrysalis.

Butterflies: Adults are easily fed in captivity in a small space. (A large flight cage is unnecessary.) The pop-up screen cage included in the kit is just nine inches in diameter and a little taller. The netting material is more important than the size. A cloth screen cage is best, because the wings begin to look tattered after a week or two in a metal screen cage or aquarium. Butterflies readily drink from fresh orange slices, but will also drink fluids such as sugar or honey water. After about ten days, the adults mate and lay eggs. The adult stage is the longest part of the life cycle. Nevertheless, even when kept in acceptable caging and fed well, the butterflies live just 21-35 days.

Mourning Cloak Butterfly
Nymphalis antiopa

Display Considerations: Butterflies should be given at least a 10-gallon net cage. Caterpillars can be kept in small containers, but require good ventilation and careful handling.

Background: This handsome, widespread species is a very special lepidopteran because the adult stage can be kept in simple displays for months rather than days. It's easy to keep the adults alive for 270 days at room temperature. They don't beat their wings to pieces in small flight cages and even hold up well with metal screen (though, cloth screen caging is preferred).

Caterpillars have branching spines that can break off in human skin and cause a sharp stinging sensation. (Generally, the spines are only large enough in the last two instars.) Usually only one or two spines enter the skin and the pain subsides in a few seconds, but handling and cage cleaning can be uncomfortable.

Husbandry: Butterflies can be provided nectar substitute or sports drinks, though I've only had

MOURNING CLOAK CHRYSALISES

MOURNING CLOAKS EMERGING

MOURNING CLOAK CATERPILLAR

MOURNING CLOAK BUTTERFLIES

good success using beetle jellies: the red brown-sugar flavored or banana flavored varieties. It is very important that the feeding cup is next to a screen cage wall so they can walk down to feed. They begin to feed three to four days after emergence, and will be seen feeding regularly if the food is accessible. If it is not placed so they can find it, they do not last many days of starvation. Daily misting with water is helpful.

As usual for butterflies, their waste looks like little paint drops that range from pink to red in color. Waste doesn't smell bad or pose any risk to the animals. A shallow layer of substrate can be kept on the bottom and stirred up now and then in lieu of daily cleaning.

The caterpillars are usually found on willow, but most willow dries out quickly even when freshly cut and placed in water. The food may have to be replaced daily as it dries out before being eaten. Rapidly withering foodplant is usually harmless, but can cause caterpillars to wander around the cage without eating much. Dry leaves can cause mature larvae to transform early, resulting in stunted adults that can be nearly as small as the preceding painted lady. The usual wingspan is 3".

TOBACCO HORNWORM SPHINX MOTH
MANDUCA SEXTA

Display: The caterpillars can be kept in a variety of containers, and accept humidity that would kill most caterpillars. Daily frass removal is still required. Adult moths make poor display animals, need very large flight cages, and rarely live more than a week.

Background: The bulky moths have a three to four inch wingspan (up to five inches in the wild). The wings are gray, while a row of six (hence the species name, *sexta*) large yellow spots runs up each side of the abdomen.

The monstrous caterpillars of this spectacular giant moth have become common as a specialty food for large insectivores. Some insect zoos keep hornworm larvae on display. This sphinx moth has been available for classroom use from biological supply houses for a few decades. The kits include artificial diet, pupation substrate (any soil), rearing containers and moth eggs.

Husbandry: When full-grown, the blue and white caterpillars are the size of a man's finger. Caterpillars are reared on artificial diet, as the volume of leaves needed to grow large numbers would make rearing a near impossible chore. If fed tobacco leaves (or any member of the nightshade

family, including potato and tomato) the blue base color changes to green. Frass should be removed daily. When mature, they begin to shrink and crawl around the bottom of the cage as they search for a place to pupate. In nature, they burrow to form an underground cell. A shallow layer of damp substrate is adequate in captivity, since the pupae don't require protection from predators. The distinct pupa looks like a bottle with a handle, and the "handle" contains the developing proboscis.

After a few weeks, the adult moths emerge and within a few days they will grow hungry. The moths normally only feed as the sun is going down and only uncoil the long proboscis in mid-flight, which requires a very large cage in captivity. Alternately, the proboscis can be carefully unrolled with a toothpick and placed in sugar water. (It takes about five minutes to feed a single moth one time.) However well fed, moths only live a week or two. Females are around thirty percent larger than males. Adults are difficult to mate. Females lay hundreds of eggs to restart the life cycle.

Though a handsome display animal, the amount of work and space required to continuously rear tobacco hornworms sets them well beyond the grasp of most keepers. It's easiest to buy them as needed. Additionally, while large batches of medium can be made cheaply with available recipes, pre-made medium is very expensive to purchase. Other drawbacks include caterpillars' constant need for attention, including frass removal and the unpleasant odor of the artificial diet (sort of like dirty diapers).

TOBACCO HORNWORM SPHINX MOTH
(TOP) CATERPILLAR
(SECOND) PUPA
(THIRD) MOTH
(BOTTOM) MOTH FEEDING

Cecropia Moth
Hyalophora cecropia

Display: Caterpillars can be kept in 2- to 30-gallon cages, a gallon or more per larva. Caterpillars require good ventilation and should not be kept crowded. Moths only live a week or so, and beat their wings to shreds after a day or two flying.

Background: Also known as the robin moth due to its monstrous size, *H. cecropia* boasts the largest wing area of the North American giant silkmoths, family Saturniidae. Females are the larger of the two sexes, but even male moths can exceed six inches in wingspan. Caterpillars are likewise gigantic and can grow to five inches in length. Though larvae are massive, they aren't the largest North American caterpillars—those of the smaller regal moth, *Citheronia regalis*, likely hold that distinction.

The night-flying female has only about five days to mate and lay 200-400 eggs. In nature, a male finds the female the first night after she emerges. Eggs hatch in ten to twelve days and caterpillars pass through five instars in about six weeks and then take a full day to create silk cocoons. Afterwards, another week or so is needed for the caterpillar to molt into the pupa.

Cecropia Moth

Cocoons of this species come in two forms. The first is the usual oval surrounded closely by an outer sheath and extensive silk anchoring that obscures the shape. In the second style the outer sheath is expanded into a large, globular, bag-like secondary cocoon that loosely encases the inner cocoon and silk lines. Wild-collected cocoons of the second form are often larger than a man's fist. The forms seem random and do not correlate with gender, population or foodplant.

Cecropia Moth Caterpillar

Husbandry: Larval enclosures should be well vented and should never be misted. Frass removal usually isn't very important during the first instars, but frass may need to be removed a few times a day in the final instar to keep down humidity and prevent caterpillars from melting. Caterpillars which die from excess humidity and limited airflow collapse (often turning black) into a pile of gelatinous goo.

Cecropia Moth Cocoon

Although most plant leaves cannot be used as food, there are quite a few choices for keeping *H. cecropia*. I've used privet, lilac, maple, sand cherry, and wild cherry. Other reported food sources include alder, apple, box elder, pecan, *Prunus*, *Pyrus* and willow. Food should be cut and placed in water to keep the leaves fresh. Leaves should be replaced when they begin to dry out.

It's difficult to rear large individuals indoors, because overcrowding and withering foodplant signal mature larvae to spin cocoons prematurely. However, indoor survival rate is 75-90% for most keepers. Sleeves (material covers) should be used for outdoor rearing, but can be circumvented by some predators and parasites. If caterpillars are placed on plants with no protection the survival rate is usually very close to zero. Considering the number of eggs produced by a female, wild survival rate is only half a percent, assuming the population is relatively stable.

PROMETHIA MOTH
CALLOSAMIA PROMETHIA

Display: Caterpillars can be kept in 2- to 30-gallon cages and require good ventilation. Moths live a week and males fly regularly.

Background: Callosamia promethia is in the giant silk moth family (Saturniidae). It belongs to the tribe Attacini that includes *Attacus, Hyalophora, Samia* and other genera. Two similar species are found in the U.S., *C. angulifera* and *C. securifera*. Promethia is a medium-size species, growing to four inches in wingspan. Cocoons and caterpillars are small for the size of the moth, so large groups can be reared with far less effort than most saturniids. Promethia's relation to the Asian atlas (*Attacus atlas*) and cynthia (*Samia cynthia*) moths is noticeable, as they likewise enjoy privet and have a relatively large wing area in relation to cocoon and caterpillar. Older caterpillars of all three have a fine, white, powdery coating. The cocoons are also structured similarly.

PROMETHIA MOTH

APPROXIMATE TIME FOR LIFE STAGES (DAYS)

Adult	7
Eggs	12
1st instar	6
2nd instar	8
3rd instar	8
4th instar	8
5th instar	8, before cocoon.
Cocoon (Pupa)	300, if kept cool.

Mating: The moths emerge best in a large screen cage, and some will mate and produce viable eggs. Some will show no interest. Hand-pairing attempted on unmated pairs is quicker and easier than with other species, but seems to only result in infertile eggs. The most successful fertilization method involves use of captive females to attract wild males for mating. Skewed emergence times are a pitfall of the abrupt temperature change caused by using a refrigerator instead of outdoor storage. However, refrigeration can be controlled.

Husbandry: First and second instars are easily kept on food plant sprigs kept fresh in water containers. As usual, make sure caterpillars can't walk down into a drowning death. Place the container in a box—the box can be only a few inches high, as long as the leaves don't hang onto or

Promethia Moth, Second Instar Caterpillars

Promethia Moth, Fourth Instar Caterpillars

close to the edge. The box should have plastic packing tape—clear or brown—around the entire inside of the top. A thin layer of petroleum jelly is spread on the tape to prevent escape—the tape is necessary because petroleum jelly spread on a rough surface such as cardboard will not prevent climbing. They normally cling tightly to the foodplant, but if the food isn't replaced before consumed or withered, the caterpillars leave. No matter how low the humidity, never mist the caterpillars, or they will die. Third instar caterpillars can be moved to a larger setup of the same type or to sleeves outdoors.

In the wild, promethia caterpillars feed on wild cherry and spicebush. It is sometimes called the "spicebush silkmoth." Alternative foodplants include birch, lilac, privet, *Prunus*, sweet gum and tulip tree. My best success has been with hybrid golden privet.

The cocoons should be kept slightly damp, as they are more prone to desiccation than other, larger species. The smaller males are the most sensitive, and it's not uncommon for only females to survive captive diapause.

Remarks: Anyone who has captured a giant silk moth at night and released it the next day, learns rapidly why they fly after the sun has gone down. Almost immediately upon release, a bird swoops in to eat it. Male promethia moths are one of the few moths that fly when the sun is still out. The smaller, dark males are believed to resemble the coloration of pipevine swallowtail butterflies, which keeps them alive as they flutter across sunlit fields late in the day. Females promethia moths are reddish-brown to pink, larger, and fly when it is dark.

SILKWORM
BOMBYX MORI

Display: Caterpillars can be kept on an open tray, and are perfect for hands-on displays. Moths are active, small, and rarely live more than a week.

Background: This sturdy creature is among the most ancient domesticated animals. The oldest English language book published on insects in 1499 is entirely about this species. The adults are short-lived, while the domesticated caterpillars are spectacular for displays and hands-on exhibits. The moths are tiny, while the caterpillars are relatively huge and impressive.

Bombyx mori is considered a separate species from its wild counterpart. There are a few different cultivars which may or may not involve crossing with wild stock. The Japanese version has nearly solid white caterpillars, adults and cocoons. The Spanish cultivar has gray caterpillars. The "tiger"-version caterpillars have dark gray bands, lemon yellow cocoons, and adults with some extra gray markings.

Husbandry: Females lay a few hundred eggs in a paper bag. This is the resting stage for this species. The eggs will change from light yellow to gray-brown within a few days and should be placed in the refrigerator. (Infertile eggs stay yellow.) They will desiccate over time and so should be kept lightly damp inside a plastic bag or a glass petri dish. A little mold is common, and won't harm the eggs, but excessive growth will kill them. After a few months the eggs can be moved to room temperature for hatching.

Hatchlings emerge over a period of days to weeks, but will emerge closer together if eggs were kept cold for six months or so. They are extremely tiny and should be kept partly covered till they molt. High humidity leads to mold and death. As mentioned, the caterpillars do become rather large, and despite the small size of the adult, will consume a volume of food equivalent to the large caterpillar size.

Moths emerge after three to four weeks. The males are smaller (and darker in some varieties), and rapidly vibrate their wings when they smell a female in the vicinity. The pair can stay coupled for days, unhooking just when the female deposits a batch of eggs.

Silkworm Cocoons and Moth

Silkworms

Domestication: In this species, the primary modifications for domestication appear to be behavioral. Caterpillars are a real joy for those familiar with rearing other lepidopterans indoors. They don't run away the minute the foodplant dries a little, and can be kept on a flat tray. The prolegs hold on tightly, but let go with ease, almost never tearing or ripping. Once they've molted a few times, they can go days between feedings with little negative impact. Of course it's hard to say how much the behavior really differs from the wild version, as its natural disposition may have always been suitable for captivity.

JUMPING BEAN CROSS SECTION

MEXICAN JUMPING BEAN MOTH
CYDIA SALTITANS

Display: A tray with edges high enough to keep the beans from jumping away. This species can't be reared in captivity, but is an interesting observational animal which requires no time or effort. It is a common curio item in some areas of the country.

Background: Most people in the United States and many around the world have heard of the extraordinary jumping beans of Mexico. However, most have never seen a real jumping bean. The "jumping bean" sold as a common novelty toy is, in reality, an inanimate piece of brightly colored plastic with a ball bearing inside. The real Mexican jumping bean is 1/2" long and brown, containing a living creature that moves on its own. The female jumping bean moth lays her eggs in the flower of a Mexican shrub. The shrub, *Sebastiana pavoniana*, grows on rocky slopes in the Mexican states of Sonora and Chihuahua. The "bean" is actually a carpel—a section of a seed capsule. It doesn't really "jump," but shakes and rocks back and forth. (Apparently, Mexican shaking carpel isn't a catchy name.) The caterpillar consumes the entire seed as it grows and leaves only the thin, hard shell. Inside this strong fortress it changes to a pupa and then exits as it molts to maturity through a circular hatch it made as a caterpillar. The bean only moves when the caterpillar is nearly mature, so the shelf life is a few weeks to a few months. Warm temperatures speed up development. Seasonal crops have been cultivated and harvested for decades, but there's no record of indoor captive breeding attempts. Interestingly, another very famous caterpillar—the proverbial "worm" in the apple—is also from the genus *Cydia*.

TRUE BUGS

Ferocious Water Bug
Abedus herberti

Display: A 5- to 20-gallon aquarium with a few inches of water and sponge islands, allowing half a dozen specimens per five gallons, assuming customary aquarium dimensions.

Background: This is an exceptionally large *Abedus* species which can exceed an inch and a half in length. It hails from the southwestern U.S. The ferocious water bug viciously attacks and sucks out the innards of insects, tadpoles and fish. This true bug is a great star for an aquatic display. *Abedus herberti* is probably the most commonly kept and bred of all the aquatic insects. A number of zoos and hobbyists maintain colonies of this energetic and robust bug. Adults can be safely kept and displayed in large groups. After reaching maturity, they live up to four years and are exceptionally hardy.

These bugs are seldom, if ever, ferocious towards their keeper. However, if carelessly grabbed near the needle-like mouth, they can inflict a bite nearly as painful as a bee sting. As with any biting animal, handling is not recommended.

Husbandry: Although *Abedus* has no ability to climb smooth glass, it readily climbs cords and tubes. The lid must be carefully made escape-proof if there are avenues of escape. Otherwise, they can be kept in shallow water without filtration as long as overfeeding is avoided. Extremely fouled water can poison them. Also, there must be plants to hang at the water's surface, or some other emergence area, or they can drown.

Captive breeding is effortless and takes place continuously at room temperature or above. (Lower temperatures or starvation cause a reproductive halt.) The female glues eggs to the male's back, where he carefully attends to them. The hatch rate is often close to 100%, but if he dies they all die with him.

Ferocious Water Bug with Eggs

Nymphs should be removed from the adult cage within a few hours, or they will be eaten. Adults can be kept together safely (a few weeks following the ultimate molt, as the exoskeleton must harden sufficiently). However, nymphs need to be reared in separate containers as they are extremely cannibalistic.

Nymphs are easily fed on tiny insects or even huge crickets and mealworms floating on the surface. Dead, sunken prey may not be noticed. If kept in separate containers with shallow water, a piece of Java moss to hang onto, and no filtration, only isopods should be used, since fouling is less of a problem. Unlike the adults, they are very sensitive to polluted water. They undergo five molts, the first four of which can take less than a week each, and the last, three weeks or so.

Display: A 5- to 20-gallon aquarium with a few inches of water and sponge islands, allowing half a dozen specimens per five gallons, assuming customary aquarium dimensions.

Background: There are only half a dozen *Nepa* species, but they are found across the globe. Members of this and some related genera look like a cross between the common waterscorpion species and the giant water bugs. They have a wide, flat body and burly forelegs similar to *Lethocerus*, but also possess the elongated breathing tubes and walking legs of the stick-like *Ranatra*. The United States hosts one small species, *Nepa apiculata*, found across the east, but it is small and difficult to acquire. Large species from Malaysia and Tanzania appear for sale now and then. Although they are similar to *Nepa* (they even have the same pattern of bumps on the pronotum), the Tanzanian species are often listed as *Laccotrephes*. Someone probably gave the Malaysian species, which is nearly identical to the Tanzanian, its own genus as well.

Wild-caught tend to be a bit beat up, but can be coated in bright green algae that make them look especially colorful. These waterscorpions have a massive crested pronotum around a tiny little head, reminding me of the end of the movie *Beetlejuice*, when the witch doctor shrinks his head. A long striped beak (rostrum) is located at the front of the head. Females are larger. The underside of the abdomen ranges from dull red to blood red. If measured "claw to tail" like true scorpions, a large adult can be over five inches.

Life Cycle and Husbandry: A 10-gallon can be set up with three inches of water, some Java moss and two 3" x 4" pieces of 2.5" thick sponge. A little Java moss is placed on the sponge and then covered with half an inch of peat from peat pellets. (Pellet peat is finely ground.) The tank is filled with water to just below the top of the sponge. Java moss aids in climbing, because the

TANZANIAN WATERSCORPION, SUBADULT

bugs can drown. The peat layer provides a media for egg laying, and the sponge keeps the peat wet while above the water level. It helps to change 30% of the water every three to four weeks.

Eggs have straw-like projections sticking out of the top. There are 2 on *Ranatra* and 5 on the Tanzanian species. Ova are buried and filaments allow them to breathe. Though egg masses can be over an inch across and 1/4" high, they aren't immediately obvious. Masses look like a pad of tiny hairs sticking from the dirt and are only visible if they are white against brown peat.

Egg masses (usually from 15-30 eggs) are laid in the peat. Unlike *Ranatra*, the egg masses are glued together at the base, like giant water bug egg masses. Egg masses are placed just above the water line on a sloping gravel bank at one end. They hatch in 30 days at 72°F. Hatchling color is a surprise. Fiery red hatchlings pop out of the top and make their way to the water. Within a few hours they transform to chocolate brown.

Though adults are safely kept in groups, nymphs must be removed and reared separately. Waterscorpion nymphs don't viciously hunt each other down like water bugs, but will kill each other as they become hungry and slowly wander in search of food.

Nymphs are moved to separate cups for rearing. They change color to match the bottom of the container. In white containers, they become translucent following the first molt, and don't turn brown again till they molt to adulthood. A bit of Java moss and ½" of water is placed in each container. Too much Java moss or too little water, and they can't molt correctly. Too much water or too little Java moss, and they drown. Daphnia can help keep the water clean, though small waterscorpions eat all the daphnia.

Waterscorpions are aquatic but live where water and land meet; drowning can be a major concern in captivity. Inability to reach the surface, or dirty water, can cause drowning. They drown if the container is filled too high. Hatchlings are the most vulnerable because they have a relatively short breathing tube. (The tube may resemble a straw, but is formed by two appendages that can be moved independently.) 100% water changes using aged tap water is necessary if the water looks syrupy. Nymphs store air under expanded flaps folded beneath the abdomen. Adults store air under the wings. Below-body air storage is why nymphs easily suffocate

Ranatra Waterscorpion from Florida

with poor water quality. Adults can survive a lot of abuse. Young grow rapidly if fed daily, but water quality suffers. Feeding once a week is safer, but never feed less often. A large container like a 2-gallon plastic shoebox for each nymph might bring die-off close to zero, since a larger volume of water is easier to keep clean.

Rearing containers used for older nymphs should be at least four inches tall. Waterscorpions can't climb the smooth sides of shorter plastic containers but every once in a while they seem

Tanzanian Waterscorpion Hatchlings

to figure out how to stand on their "tail." They jump containers and kill other nymphs, or end up on the floor and dry up. Escapees may seem to survive a day on the floor. When found and put back in water they look okay, but are lifeless the next day.

Hatchlings can be fed prey larger than themselves. *Nepa* and *Laccotrephes* are more aggressive than *Ranatra*. Waterscorpions can be started with fruit flies, terrestrial isopods, mealworms, daphnia, crickets, cockroaches, meal moth caterpillars, etc. Large nymphs and adults capture tadpoles and small fish. They'll drink from gigantic prey but the leftover carcass should be removed.

Nymphs molt a total of five times after hatching to reach adulthood. The first molts take as little as a week, though the final molt takes a minimum of one month. Prior to the last molt, wingpads turn tan, then black, and in three or four days the adult emerges.

In captivity, giant Tanzanian waterscorpions lay eggs year-round and incredible numbers can be reared in a short period. However it can take time to work out all the bugs. Females push peat into the water, so it has to be replaced regularly. With minimal feeding, months can pass without eggs. At times the females drop all the eggs in the water. (Eggs are dropped if the water level gets low, the substrate is the wrong type or there is not enough of it, or there are too many plants or decorations in the water.) Eggs do not develop correctly when the water level is much below them, even though they don't dry out. Cannibalism takes its toll, because hatching occurs in large groups or not at all. Small containers (due to space limitations) can lead to high losses.

Wild adults are generally easy to sex as females are larger. Nevertheless, captive-reared adults can't be sexed using size. The underside of a mature female's abdomen becomes rounded within a few weeks, while the male's remains keeled. Captive-bred adults live two to four years.

Waterscorpions don't attack fingers placed in the water, and hold their bodies stiff when picked up. They are easily handled without trouble. This family is said to lack the stink glands common to many true bugs, but they can produce a smelly, white defensive fluid. A number of the predatory, aquatic true bugs can inflict painful bites. However, the only way to see if waterscorpions can bite is to purposefully force the rostrum into flesh. They seem incredibly meek, but throw in small, energetic prey, and they come to life.

WHITE SPOT ASSASSIN

WHITE SPOT ASSASSIN
PLATYMERIS BIGUTTATA

Display: A 5- to 20-gallon aquarium with an inch of soil or coco fiber. Display colonies are kept crowded, but nymphs can be reared separately at first. They hide under logs in nature, so cork leaned against the front and lack of other hides is important. Caging should be secured from human access. The bite is very painful.

Background: Platymeris biguttata are found under logs where they feed on ground dwelling invertebrates. This and a few related species are found across forested areas of Africa. The similar red spot assassin, *Platymeris rhadamanthus*, averages smaller, is slightly less robust, and apt to ruffle its wings when disturbed. Specimens are not aggressive towards keepers, and do not bite if touched or held, but most certainly bite when grabbed or restrained. The venom causes extreme pain. There is no record of allergic reaction, but there is little information. They are capable of spitting venom, but rarely more than a quarter of an inch. The ability, however, is advanced in

RED SPOT ASSASSIN LIFE CYCLE

WHITE SPOT ASSASSIN

the larger *Platymeris* sp. "Mombo," which can spit venom up to two inches. Also known as the orange spot assassin, the original stock was collected in the small town of Mombo in Tanzania.

Husbandry and Life Cycle: The large African assassin bugs are a common display animal, since they are large and easy to care for. Adults are brightly colored and can be kept safely in huge groups. They feed on crickets, but will also eat cockroaches, isopods, and other small invertebrates. Rearing cages for immatures should contain paper egg carton rather than decorative bark, since the increased area greatly improves survival. Once the colony reaches a good size, a high survival rate is no longer a good idea and nymphs can be left with adults.

ORANGE SPOT "MOMBO" ASSASSIN

Adults look similar from above, but from below the genders are easy to discern. The male's abdomen ends in a cup-like segment covering the genitalia. As they are kept in colonies, fertility is not an issue as long as both genders are present. Females lay small brown to black eggs with a bright white operculum (or lid). These can be buried in shallow pits in damp areas of the cage or dropped below the adult's hide spot. Ova should be kept damp, but will survive limited dry periods and can even recover from temporary dimpling (where the egg shell contracts inwards from loss of moisture). Colonies produce well with all stages kept and fed together, but initial startups can take some time, space and effort due to cannibalism.

MILLIPEDES

African Giant Black Millipede
Archispirostreptus gigas

Display: 5- to 30-gallon cage with damp substrate and limited ventilation. Numerous adults can be kept together without aggression. Substrate should be shallow to prevent burrowing.

Background: Their sleek, powerful appearance impresses even the non-invert fan. It would be difficult to find a person who hasn't seen or heard of them. They've been at pet shops for as long as anyone can remember, but are rare now as they are no longer imported. They survive a certain amount of overcrowding and abuse, which is the hallmark of a good subject.

The common name is descriptive. They come from Africa and are gigantic. An exceptional adult specimen can measure ten inches, though seven to nine is typical. Even a seven inch AGB appears to be a foot long. The body is as thick as a cigar. The adult's base color is deep, satiny black. Some are brown along the underside, the legs or between the segments. The scientific name is *Archispirostreptus gigas*.

The powerful, gentle feel of an AGB's grip as it walks across your arm is strangely compelling. Because they are so easily handled, AGBs have been extremely popular for handling shows and displays. They are not venomous and are unable to bite, pinch or scratch. Still, they aren't completely defenseless. Frightened millipedes discharge a foul-tasting, yellow fluid—the primary active ingredients are organic acids (quinones). The defensive fluid causes a temporary skin discoloration of yellow and then red-brown. The fluid is painless unless it is rubbed in eyes or an open cut.

Life Cycle and Husbandry: Males tend to be smaller than females, but size is a poor indicator of gender. It's easy to determine gender with 100% accuracy by looking at the legs on the seventh segment from the head. These legs appear to be missing on the male because they are modified into gonopods (secondary reproductive organs that resemble "white bags" when exposed), used during mating. A mating pair looks like they are hugging.

Females deposit 1/16" light yellow eggs a few inches deep in substrate. They don't construct the familiar spirobolid egg capsules nor build nests like millipedes from other orders. As common for spirostreptid millipedes, eggs are laid bare in the substrate and are far more durable than the delicate eggs of most millipedes. Nevertheless, eggs can be smashed. They should be removed from the adults' cage for the best survival rate. (Eggs are tough, but early larval stages and molting immatures are easily damaged.) Eggs are laid one at a time, but over one or two months each female can produce more than five hundred eggs.

The substrate must be damp for egg laying and can be as shallow as 2", but 4" to 6" depth is recommended. Potting soil, coconut fiber, rotten wood and compost are all acceptable substrates. Females are picky and will not drop eggs just

Immature African Giant Black Millipedes

anywhere. They refuse to lay eggs—or lay just a couple—if there isn't a stimulus. Correct level of decay, type and volume of decayed leaves is the stimulus. An inch of old leaves added to damp substrate is what is needed to encourage a healthy adult female to begin egg laying. Choose decaying leaves (a mix including oak and beech) that have been on the ground a few months or more. Avoid very dark brown to black leaves or leaves that have turned yellow from fungus. Thoroughly heat leaves in the microwave oven to prevent the introduction of pests such as earthworms and nematodes. Keep in mind, egg laying takes place only once a year and the best time is August, although captive-reared specimens might lay eggs at any time.

The egg transforms to a protonymph in a month. Although the protonymph is bright white, the yellow-white egg shell often darkens to red-brown during incubation. The protonymph resembles a tiny croissant and is motionless, legless, and without visible ocelli. It is more vulnerable than the egg and easily trampled by adults. After a month, the protonymph hatches into a tiny white millipede with legs only on the first half dozen segments. Though it has very few legs, it burrows nimbly into the substrate to feed. In three to four weeks it molts and gains legs on all but the last few segments. A quarter of the eggs make it to this point, but females lay countless numbers, so the survivor count is generally still in the hundreds. Die-off after this stage is minimal.

Six months following hatching, the millipede is light-gray and only 3/4″ long. It grows a little darker at each molt. At two years it will be dark brown and three to four inches. Immatures molt every few months, but can take up to a year for the last molt. Sexually mature AGBs rarely, if ever, molt.

Molting is a process. The millipede builds a large cell and the body turns grayish. Within a few weeks of building the cell, the animal molts and the old exoskeleton splits dorsally from head to tail. The shed is thin, frail and off-white. It will be eaten unless the AGB is disturbed. The animal is extremely vulnerable during the molt, far more delicate than tissue paper. Overcrowding or handling leads to damage and death. Minor damage heals at every molt but is still visible many molts later. The exoskeleton grows stronger over the next few days, until it again feels hard as a rock.

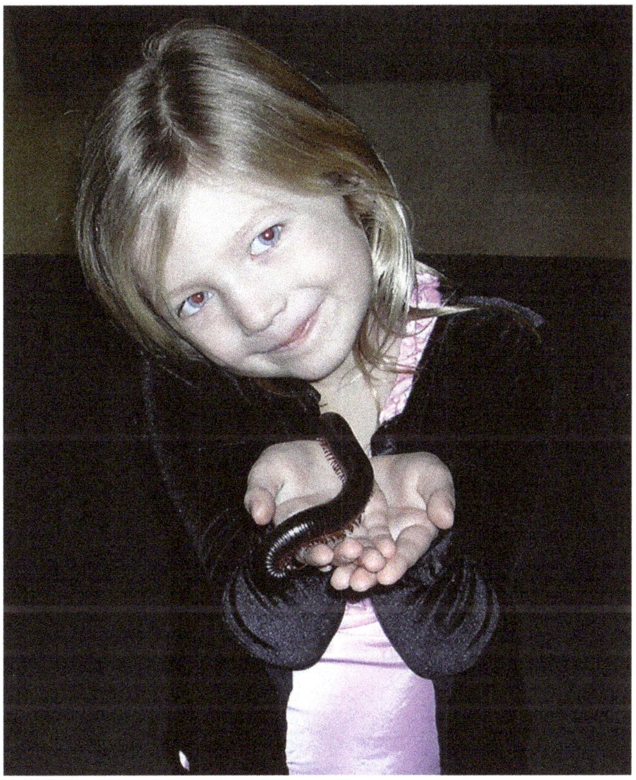

African Giant Black Millipede

AGBs feed on a variety of human and pet foods (dog, cat, fish, etc.) and most fruits and vegetables. Watermelon rinds and pumpkin are a favorite. All these mold and rot quickly, so uneaten food should be removed after 24 hours. Edible substrates like rotten wood and old decayed leaves aren't volatile. They can be kept in the cage until eaten and keep millipedes healthy for months or years without additional feeding, though growth is slower. Keep in mind the rotting wood has to be very soft or finely ground for a millipede.

AGBs take three to five years to reach adulthood and live a decade or more if properly cared for. They can survive months, but seldom years, in substandard caging where stagnant or overly dry conditions cut their lives short.

Keys to Longevity: The cage should contain a minimum two inches of damp substrate.

Ventilation is required and should not be excessive or insufficient. A solid glass lid on an aquarium should have a 1/4" gap along one side. A plastic shoebox should have a dozen 1/8" holes drilled in the lid. Ventilation can be increased if daily attention is paid to maintaining the moisture level.

The surface of the substrate should not be allowed to dry.

Avoid overfeeding, because excess rotting food can poison the air and kill the millipede.

GLOSSY BLACK PINKLEG MILLIPEDE
DENDROSTREPTUS MACRACANTHUS

Display: 5- to 10-gallon cage with damp, shallow substrate and limited ventilation. Thick, rough climbing branches can be offered. Very defensive, this is a very poor choice for handling displays.

Background: This huge African species, found in Tanzania, is one of the few brightly colored spirostreptid millipedes and one of the largest species successfully bred and reared in captivity. The body from head to telson is glossy black with a strong resemblance to patent leather. (It glows dark forest green under black light.) The legs are long in relation to the body size and are a delicate pastel pink with lighter pink banding. Females grow to eight inches, making them the longest of any commonly kept species other than the familiar African giant black. Nevertheless, the body is thin, so a couple of shorter, thicker species are more massive. Males are seldom over six inches. Members of the family Spirostreptidae are often especially long and thin, and include *Macrolenostreptus*, *Orthoporus*, as well as the familiar AGB—*Archispirostreptus gigas*.

GLOSSY BLACK PINKLEG MILLIPEDE AND EGGS

Reports indicate *D. macracanthus* are primarily arboreal in the wild and are almost never encountered at ground level. The elongated legs and bright colors can be easily attributed to life in the trees. They also have two behavioral traits that may coincide with an arboreal nature. First, they readily secrete much more defensive fluid than other millipedes, which would be needed protection for a constantly exposed animal. Second, all the eggs are laid in a day, in contrast to the weeks to months most millipedes spend ovipositing. Despite arboreal habits, captive caging elements used for terrestrial spirostreptids work well, and immatures build underground chambers for molting. (Adults of this species do not molt.)

This millipede does not appear to be long-lived for a giant millipede, since adults are commonly dead in three years or less. A few have lived longer, so it's possible with the four to six years required to reach maturity an exceptional specimen might live a total of ten years. The lifespan is gender biased, as males usually die first.

The primary millipede foods of rotten wood and dead leaves are readily eaten. Additional foods like dry dog food and most fruits—a favorite is cantaloupe rind—can be offered regularly, but should be removed if uneaten after a day or two.

Husbandry and Life Cycle: Adults are easily sexed from above because males have expanded segments from the head back to segment seven. Also, females are much larger. Both genders are the same color at all stages, so immatures can't be sexed until they reach a few inches, when the presence (male) or lack (female) of gonopods can be checked.

Adults can be set up in a leaf and compost mix used for spirostreptids. The footprint of the cage (length times width) should be at least 9″ x 14″ while cage height can be anywhere from 4″. Total cage venting should be not more than 10% of the footprint surface area—at minimum, just over a square inch. Substrate mix depth should be 2-5″ with an additional 1″ layer of rotten old leaves—on the ground at least one winter—on top. Leaves should be carefully cooked to prevent the introduction of pests that eat the food, proliferate, and kill immature millipedes. Some keepers choose to freeze the substrate instead, but few temperate pests can be killed with cold.

Glossy Black Pinkleg Millipede

Males don't seem to fight over females, but mating isn't seen that often and generally occurs in the late evening. During mating, the male rhythmically releases and closes all of his legs along his entire body, about once every two seconds. The leg-pulsing behavior doesn't serve an obvious function but has been observed in some other species, including *Epibolus pulchripes* (a huge spirobolid also found in Tanzania and Kenya).

About once a year the female builds a loose, shallow cavern and deposits a few hundred yellowish, 1.5mm eggs in a large, slightly sticky mass (they appear stuck together, but a slight touch and the mass collapses). Eggs are hardy and similar to *Archispirostreptus* eggs, but only about 2/3 the size. Females practice maternal care and remain through second instar, but do not return if disturbed. This is the only giant millipede bred so far that lays bare eggs in a single, large group rather than spread out. (*Pachydesmus*, a flat millipede, lays a large mass of eggs inside a frass structure.) Females housed together produce batches of eggs within a month of each other. A mix of primarily oak and beech leaves, which had been on the ground at least half a year, was used to induce egg production one year. The second year, maple leaves that had been collected in fall, and stored damp in a trash bag for two years, were used. The number of eggs laid the second time were minimal, but the low numbers may have been caused by age or eggs were possibly destroyed by the immatures still present from the previous year. Protonymphs hatch from the eggs after a few weeks and molt into white immatures with only a dozen legs, six at the front of each side. At the next molt, they have legs on most segments but don't have the maximum number of segments yet. As they grow and add segments, they darken over a number of molts and transform from white to pale, gray, brown and finally black at two to three inches. They look very glossy even when very tiny. Young millipedes are hardy and don't seem to experience the usual, noticeable die-off resulting from substandard food. They grow at a good pace and some reach two inches within nine months of the eggs being laid.

Thai Rainbow Millipede
Apeuthes sp.

Display: 2- to 20-gallon cage with damp, shallow substrate and limited ventilation.

Background: This appealing animal originates from Thailand and has been exclusively available as captive-bred specimens. It is a colorful spirobolid millipede, though its name comes from a prettier, poorly-adapted for captivity, close relative commonly known as the Vietnamese rainbow. The red-margined, steely blue segments of the Thai have a truncate yellowish band at the front, which is only visible when the segments bend. The telson ends in a pronounced, rounded point. Adults normally grow to four inches, although they continue to molt after sexual maturity and can grow to 5″. It's difficult to estimate longevity, but I have at least one adult I received as an adult 14 years ago, most of its siblings died a few years back. Specimens usually don't emit defensive secretions during handling, which is a very pleasant characteristic.

Husbandry and Life Cycle: Though easy to keep and breed, availability is kept in check by a long life cycle. Immatures are seldom longer than an inch after a year and many molts. Three to six years are required to reach maturity. Even after they are sexually mature, it can take a number of additional years to get good egg production out of the females. In one group (9 males, 11 females), one young adult female made hundreds of empty egg capsules eighteen months after reaching maturity, another laid a few dozen viable eggs, while the remaining adult females mated but did not produce eggs for the first twenty four months. Egg laying requirements are the usual for spirobolids and eggs can be retained indefinitely if the habitat is far outside the normal range. Eggs are usually laid over a period of thirty to sixty days, once a year, and one female should lay hundreds of eggs each season. The female carefully forms a large capsule for each egg that looks similar to a frass pellet. If the protective capsule is broken, the egg dies.

Eggs develop into protonymphs and then molt to first instars that remain inside the capsules. They usually do not leave until second instar, at which point there will be hundreds of tiny, colorless rainbows in the cage. Immatures are white for a number of instars.

THAI RAINBOW MILLIPEDES

Young may survive neglect that kills most other species, but care should be taken with these gems. Replace the substrate when it becomes mostly frass, and offer melon rinds, dog food, pumpkin and rotten leaves on a regular basis.

ALBINO AMERICAN MILLIPEDE
NARCEUS AMERICANUS

Display: 1- to 30-gallon cage with shallow, damp substrate and limited ventilation. Numerous specimens can be kept together.

Background: Narceus americanus is a large species of spirobolid millipede found across much of the central and eastern United States and southern Canada. At three and a half to five inches, it is the longest native invertebrate in the northern areas of its range. Adults vary from a steely blue to grayish-brown base color accented with red to pink legs and segment margins. This section concerns the captive reproduction of the rare albino form of this species.

There are a few members of the genus *Narceus* in North America, including *N. gordanus* and *N. annularis.* An adult male specimen was sent to a taxonomist, Dr. Richard Hoffman, who confirmed the *N. americanus* species identification.

Reproduction: The original pair were the only two albino specimens from a large group of captive hatched and reared normal-form *N. americanus*. Egg capsules from the albino pair were first observed on June 15th of 2005. This was a relatively small batch for *Narceus*, as only forty-eight immatures resulted, but it was still a thrilling achievement. Of those, twelve were normal coloration and thirty-six albino. The second batch of egg capsules were formed in late June 2006. Approximately 450 immatures resulted from this batch. Only 2 were normal coloration.

Adult Molts and Lifespan: The female molted once each year around April, while the male did not undergo any post-maturity molts in my care. (*Narceus* males commonly go through post-maturity molts, but not with the regularity of females.) Unlike some spirobolids, *Narceus* are not sexually mature before reaching approximate adult size, and post-maturity molts are not accompanied by increases in mass. The male died in February of 2007, and the female died in December 2007. The adults were almost certainly siblings and likely hatched out within a few weeks of each other. They had probably been mature for less than two years before I received them in 2004. The lifespan might be much longer if provided natural cold cycles, but these specimens were never exposed to cold temperatures. The total lifespan at 70-76°F appears to be only six or seven years, which is short compared to *Narceus annularis*, which can take five years just to reach adulthood at room temperature and continues on much longer.

The Subsequent Generation: Ten females and seven males from '05 were retained for breeding. Eggs and 2nd instar were observed on October 2, 2007, from the second generation adults. Some were eggs and some 2nd instars, so egg-laying probably began in late August. This was much faster

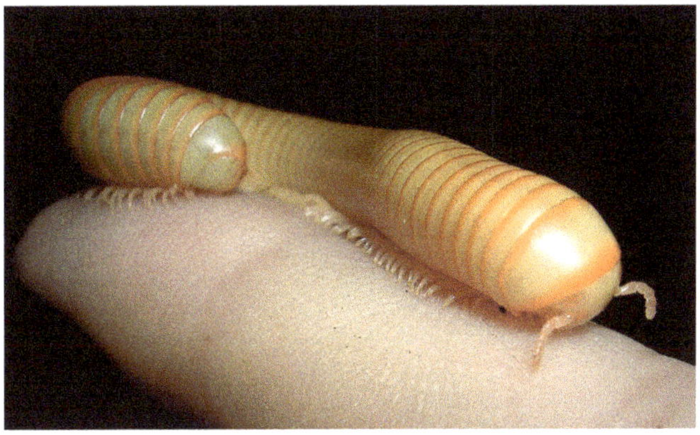

ALBINO AMERICAN MILLIPEDE

than I would have imagined. The *Narceus annularis* I had reared for years do not mature in less than three years, and don't lay eggs for the first year after they reach maturity and are able to mate. From the number of young produced (36) it's fairly certain only one of the ten adult albino females laid eggs.

With the large number of adults, the second breeding cage was a 10-gallon aquarium with a solid sheet of glass on top. Excess silicone was removed from the corners, but the adults still have some ability to climb the corners. The glass is not airtight, but prevents escape and maintains high humidity.

The Albino Trait: Albinism is generally considered a lack of ability to produce the primary dark pigment (melanin in many animals), not an absolute lack of pigment. Like vertebrate albinos, albino *Narceus* are not completely devoid of yellow or red pigments, just dark pigment. Yellow and light red are maintained, but are only notable in older immatures and adults (though the ozopores of small immatures appear red). While some animals can produce a pied or gradient albino, the albinism in *Narceus americanus* for around a thousand individuals produced over four generations has been 100%.

After the first batch resulted in 25% normal, 75% albino, it seemed the gene for albino was somehow dominant, which didn't make sense. However, the second group resulted in less than

0.5% normal. The female had been with normal males prior to arrival, but had molted since and it had been more than a year. The most likely conclusion from these numbers is that sperm retention in the female can last through molts and the albino trait is in fact recessive. 100% of the subsequent generation were albino (as females were never exposed to normal males).

Remarks: Although much of their time is spent concealed in nature, the albino form is probably quickly removed by predators on emergence. It's fantastic to have a captive-produced invertebrate variant, but what's even better is the albino *N. americanus* is extremely attractive and hardy. Unlike the regular form, the handsome color pattern is unique among millipedes. The recessive albino trait has had no negative effect on health or survival. The albinos have one of the highest survival rates for any millipede I've worked with.

NORTHEASTERN GIANT MILLIPEDE
NARCEUS ANNULARIS

Display: 1- to 30-gallon cage with damp substrate and limited ventilation. Numerous specimens can be kept together.

Background: This is one of the largest and most colorful North American millipedes. The legs, antennae and margins of each body segment are pink to red. The base color of body segments is dark bluish gray. Adult females reach five inches, and their body is roughly the diameter of a pencil. Males are thirty percent smaller.

Adult males are easy to differentiate from immature females without flipping them over to check for missing or modified legs on the seventh segment. The dorsal surface of the sixth and seventh segment is visibly widened. These modified segments are located about half an inch back from the head.

Narceus annularis is found across central and eastern United States and Canada. It is a common species but most hobbyists have never seen it and are quite surprised to discover that such a large millipede is found in their area. Like the majority of millipedes, it is nocturnal. Only occasionally does *N. annularis* wander from its feeding grounds.

Collecting: Mass exodus of *N. annularis* is an uncommon occurrence. Groups of hundreds and hundreds of adults have been observed fleeing from part of the forest (J.T. Bowman, *pers. comm.*). Where they are going and what prompts this behavior (i.e. season, weather, lack of food, etc.) is a mystery. If the hobbyist is lucky enough to run into a column of hundreds of fleeing millipedes, collection is easy.

However, the collector seldom gets so lucky. Normally, eastern giant millipedes have to be carefully hunted. Finding one or two specimens during a few hours of collecting is good. Rotten logs or trees pocketed with cracks full of rotting leaves are prime habitat. The quantity of partly rotted trees and logs is more important than the size of the wooded area. Bear in mind, feeding areas are seldom at ground level, usually 1 to 10 feet above ground.

Millipede frass is the most obvious indication of nearby giant millipedes. Frass is roughly the size and shape of a bb. The surface of the frass is coarse and tends to be the same color as the surrounding food. When a piece of bark or wood is pulled away, hundreds of the little spheres can be seen. Each millipede produces tons of the small objects and it is common to find only one millipede for half a gallon or more of the frass. They tend to be located inside or within six inches of frass deposits. Usually, more frass equals more millipedes. However, sometimes a lot of millipede frass is found over a large area and even though there appears to be more than enough food in the area, there are no millipedes to be found. Possibly a mass exodus of *N. annularis* is responsible for the absent animals?

NORTHEASTERN GIANT MILLIPEDES

Feeding and Caging: Narceus annularis consume large quantities of mushy, rotten wood and decomposing leaves in nature. It is important to include decaying wood and leaves in the captive diet. Supplemental foods include dog food, fish flakes, fruits and vegetables. Adults can survive for years on dog food and fruit, but do not thrive and are unlikely to breed without an appropriate diet. They do best with a variety of foods.

A common problem associated with supplemental foods is mites. Uneaten dog food, etc., allows mites to multiply out of control. Remove uneaten food after twenty-four hours to prevent this problem. If a mite infestation in noticed, replace the substrate to manually remove the majority of mites. Then, feed the millipedes only with decayed wood and leaves to eliminate the mites' food source.

Centipedes eat millipedes. Full-grown eastern giant millipedes are far too large to be eaten by anything other than the giant centipedes (*Scolopendra*). On the other hand, young millipedes are easy prey for even the smallest native centipedes. Decaying wood and leaves almost always contain centipedes. Remove centipedes by heating or grinding collected food.

Large *N. annularis* are hardy and can endure desiccation or substandard ventilation for many months. Still, millipedes kept in poor cages die prematurely. The cage should conserve moisture and provide good ventilation. Use a plastic shoebox with twenty 1/8-inch holes or an aquarium with 90-95% of the lid covered. Place an inch or two of moist peat, potting soil or finely ground wood or leaves on the bottom. Deep substrate is fine, but unnecessary because the millipedes rapidly add to it as they eat.

Breeding: Well-fed, healthy adults breed readily in captivity. Multiple males and females can be kept together without rivalry. During mating, the pair "hugs" and the two appear joined at the seventh segment from the head. After an hour, they separate. In nature, the male and female of a mated pair are often found together.

The female forms hollow dirt spheres and glues a single egg to the inner wall. She normally constructs 200-300 of these capsules over a four week period. In the wild, she lays eggs each year around mid-June, but can produce a second batch in December in captivity. The egg capsules are nearly indistinguishable from the frass, but are slightly more symmetric and have a consistent texture. It is important to carefully check frass to avoid throwing out eggs or hatchlings.

Growth: Young millipedes begin to appear about a month after placing breeding adults in new substrate. Due to the cryptic nature of egg laying, it is difficult to determine the exact gestation of the eggs. Eggs and hatchlings usually die if the capsule is opened. The pale, fragile larvae molts before leaving the egg capsule. The young millipedes resemble the adults, but are grayish-white and have only six legs. Another two to three molts are required before the young approximate adult coloration.

In deep, moist substrate, millipedes dig down and construct a molting cell. Inside the cell, the new exoskeleton forms, the old shell is shed and then the new carapace hardens. In loose or shallow substrate, they cannot form a cell and molt just as well on the surface. Be very careful if digging through deep substrate, as freshly molted millipedes are easily damaged or killed. If a millipede appears white or chalky, place it on the surface. If reburied, it may not have the strength or time to construct a new cell.

Narceus annularis seldom chemically defend themselves unless severely mistreated. However, freshly molted individuals freely secrete chemicals known as quinones that stains human skin yellow, later changing to red then brown. It does not wash off and can take over a week to fade away. The defensive fluid has a caustic odor and stings if it gets in your eyes. The color of the new exoskeleton is noticeably deeper and darker. Avoid handling freshly molted millipedes.

This millipede matures slowly. At room temperature, this species takes at least three years to reach adulthood. In contrast, captive-hatched *Chicobolus spinigerus* (a related, slightly smaller species found in the southeastern U.S.) mature in one year.

Difficulties: Wild-caught adults may average a higher percent of females (sample population from northern Ohio contained 4 males: 7 females). However, the first batch, laid in September of 1995 and maturing in late 1998 to early 1999, resulted in 31 adult females and 0 males. The female was caught in May of 1995 in Ohio and the male in late August of 1995 in South Carolina. The male was most likely *Narceus americanus*. The original female did not produce eggs until after mating, and the captive-reared females did not produce eggs until males were acquired. Parthenogenesis due to incompatible sperm is the first of many possibilities. About a hundred had been traded off as medium and large young, and fifty or so were lost at various stages. All the males might have been traded off accidentally because of size differences. Due to the belief that with over thirty left, there had to be some of each sex, there wasn't an effort to determine the sex before trading. It is also possible the females may have been hardier, and most males died before maturing. Another possible influence is the temperature of the eggs during gestation. Egg sites encountered in nature are in cool locations inside hollows in trees and logs. Warmer room temperature gestation could have caused the eggs to develop into females.

(Top Left) Chicobolus Millipedes, Immature
(Top Right) *Chicobolus* Millipedes, Ebony Form
(Bottom Left) *Chicobolus* Millipede
(Bottom Right) Chicobolus Millipedes, 1st and 2nd Instars

CENTIPEDES

Scolopendra heros

Display: 5- to 10-gallon cage (or 1- to 2-gallon jar with screw-top lid) with damp substrate and limited ventilation. One specimen per cage.

Background: Scolopendra heros is the largest North American centipede, with three separate and spectacular color phases. Historically, the three have been considered subspecies, though the geographic ranges are not well-known. These variants or races noted by Shelley (2002) as "color pattern(s) traditionally assigned to the variant named ..." which seems a roundabout way of recognizing the uniqueness in an attempt to avoid it. The stark contrast in forms without any seeming intermediate forms raises the question whether they might even prove to be different species were DNA analysis ever performed.

Husbandry: Care is generally straightforward, and specimens tend to live at least a few years. However, there is some die-off of very small or captive hatched young. Moisture is very important, but a section of the substrate can be kept dry to allow the animal to regulate its moisture. Large specimens may be kept alive for a time with only a water dish and dry substrate, but usually not for too many months. The problem with offering damp substrate is that it is very easy to overfeed centipedes. Wet, rotting insect parts leads to grain mite infestations and health problems. Centipedes do not need to eat much, and if the specimen is leaving behind scraps, the refuse should be removed and the feeding frequency reduced. A large plastic jar with screw-on lid is the recommended cage as it is escape-proof when closed. Six to twelve 1/16" holes should be drilled in the lid to provide ventilation.

Subspecies:

Scolopendra heros arizonensis

Consistent coloration: Last three tergites (dorsal body segments) black; bright orange terminals and antennae; head dark; legs whitish tan.

Common variations: Two primary variations, banded (normal) or unbanded. 1st tergite is usually black on unbanded specimens, tan on banded. Body coloration ranges from tan to reddish-orange; reddish-orange coloration usually only on small, young specimens.

Range: Arizona, except for northwestern and extreme western regions, and the southwestern "boot heel" of New Mexico (Animas and Peloncillo Mountains in Hidalgo County).

Scolopendra heros castaneiceps

Consistent coloration: Orange head and first two tergites; dark body; light orange legs, though the terminal pair is the color of the body, and transitions to orange at the apex (tip).

Common variations: Orange can vary from light orange to reddish orange. Adults are greenish to

Scolopendra heros castaneiceps

Scolopendra heros arizonensis

brownish black. The bodies of some older specimens fade to greenish brown. In many hundreds of 2nd instar specimens seen in captivity, the main body starts out bluish black.
Range: Northeastern New Mexico, all of Texas except the southwest, east through western Arkansas and Louisiana, north to Kansas, and southwestern Missouri. Colorado is commonly listed, and the southeastern tip follows the likely range, but there aren't any verifiable reports available. Attems includes Georgia and Alabama (1930), but there are no verifiable reports.

Scolopendra heros heros
Consistent coloration: Light, golden tan body; lighter legs; tan antennae; last two tergites are black.
Common variations: Banded specimens have a dark head. Unbanded (normal) have a light colored head. Terminal legs are usually solid tan on banded specimens; these are dark at the base, transitioning to tan, on unbanded specimens.
Range: Southwestern Texas to south central New Mexico north and east of "boot heel" region, also northwestern Mexico.

Minor banding and color variations within subspecies may cause some confusion, but overall color patterns still hold true. In addition, the exact shade of a color is influenced by the age of the animal. Furthermore, color depth, not pattern, can change when reared under captive conditions as a result of nutrition or environment. There was a fourth variant described, *prismatica*, by Cragin in 1885, but there is no type specimen, and nobody has seen one since 1885.

The occurrence of all subspecies of *Scolopendra heros* is usually associated with rocky areas. Rarely, it is found in east Texas, usually in rocky outcrops that are not common in the region.

Specimens are commonly found within walking distance of water sources.

Scolopendra heros and Scolopendra polymorpha confusion: The only species that seems to be confused with *S. heros* is *S. polymorpha*, but there are easily observed differences. A unique feature of *S. heros* is that the second maxillary palps are toothed rather than tapered, but this feature is rarely visible in photographs and is difficult to see on live specimens. One feature is commonly visible in photographs and in person: *S. heros* usually have 3-5 smooth basal antennae segments whereas *S. polymorpha* normally have 7 or more.

The reason for the confusion is that both can grow large and are found in many of the same areas. While some *S. polymorpha* can be sexually mature at 1.5", others are capable of exceeding 5", which is the average size of adult *S. heros* (which range from 4-7" at maturity). There are reports of solid yellow/gold *S. heros* from south Texas, but the few specimens I've seen labeled "gold" *heros* were *S. polymorpha*.

Feathertail Centipede
Alipes grandidieri

Display: 1- to 5-gallon cage with damp, shallow substrate and limited ventilation. Specimens are sometimes kept in groups successfully, and other times they eat each other. They make rasping noises with the rear legs, but it's difficult to get them to stridulate on command.

Background: The most unique giant centipedes are without a doubt the noisy species in the genus *Alipes*. They look like normal Scolopendrids, except the rear legs are highly developed and expanded into paddles with feather-like markings. The bizarre rear legs are used to make squeaky sounds likely used as a warning to predators. Though they are not aggressive, they can inflict a painful bite. In captivity their first choice is to make noises when harassed. The male and female have the same level of "tail" development, so it is not likely to be important to courtship. The tails are shaken outwards at a ninety-degree angle for stridulation, not rubbed together. Animals with only one rear leg make the identical sound, but at only about a third of the volume.

In the wild *Alipes* are found along the shores of African lakes, including Lake Victoria. They are found under stones and are solitary. *Alipes* move with the receding shoreline that results from seasonal drops in the lake levels. The shore often recedes twenty feet during the dry season.

The feathertails described here came from Tanzania through an importer in California. The centipedes were marked down 75%, so I bought a dozen. The lowered price may have been because the feathertails arrived missing one or both tails. The worst was missing eight legs, and the best missing two. They were also only half grown, which wasn't a bad thing. They were more apt to grow back the legs. Adult centipede molts take place at greater intervals, and some types may not molt in adulthood.

Locations of Alipes (according to Attems 1930):
Alipes appendiculatus: Zambia, Zaire, Uganda
Alipes calcipes: Angola, Congo, Zaire
Alipes crotalus: South Africa
Alipes grandidieri: East Africa, Tanzania, Zanzibar
Alipes multicostis: Cameroon, Nigeria, along the Gold Coast to Sierra Leone

Reproduction: Most looked almost normal in a single molt, but one animal took four molts to grow back both tails correctly. None died from their injuries, though one was dried out and dead on arrival. (Its container had very obviously been overlooked, as the paper towel was bone-dry.) Each underwent three to five molts that occurred in the first eighteen months, and none in the following twelve months. At the end of two and a half years I had kept three. Then, the biggest one died. Since there hadn't been changes in care and it hadn't molted in a longer time than the others, the most likely cause was old age.

FEATHERTAIL CENTIPEDE, PAIR

The last two looked identical, but none of the feathertails had displayed any sexual dimorphism. I decided to put the two together; if they were approaching the end of their lives, they deserved a shot at reproduction. More aggressive centipede species normally don't attack each other right away if given space, media to dig in, and food. However, the risk of one being eaten increases with every passing night.

Alipes adults are under four inches, trunk length, so a 5-gallon aquarium was chosen. Setup included 1/2" of aquarium gravel covered with 2" of ground coconut shell. The substrate was kept very damp, since even adult feathertails are vulnerable to desiccation. The lid consisted of a piece of glass with a 1 1/2" square opening at one corner, and a screen lid over top. Decorations were a section of corkbark and a length of driftwood. They crawled away to opposite corners after touching. I checked on them a few days later, and found they had carved out a shallow under the driftwood where they were together! Other species tend to have separate burrows dug in opposite corners after the first day or two. They were given four or five 1/2" crickets twice a week, and pillbugs were placed in the tank for cleanup. Some were eaten.

They lived together four months before she ate him. It's not possible to say she killed him. He may have died of old age, though he had not shown any signs. (Commonly, centipedes look dull or sluggish and lose their appetite weeks or months before death.) Whether he died from love, age or something else, she did eat him.

Less than two weeks later she was holding eggs! Approximately forty days later, when the eggs developed into protonymphs, it appeared she had eaten a mate and not just a close friend. Protonymphs are a lot bigger than the eggs and in this genus they resemble a tire. (*Scolopendra*, *Hemiscolopendra* and others tend to look like croissants at this stage.) Progress was checked every week in order to take photos and gather development information. Each check was a relief because she hadn't eaten the immatures, but worrisome since this time could be the disturbance to upset her. Each time the female was carefully nudged back into position under the driftwood and substrate placed over the sides of the hollow to seal out light and intruders, and keep in humidity. Following thirty days, the protonymphs hatched into first instars, which look like normal centipedes, except they are yellowish-white and don't eat. The female holds onto first instars until they molt, as is common for Scolopendrida. (In other centipedes the first instar is solitary, fully mobile, and/or can feed.)

Less than twenty days after the change to first instar, tiny antennae were waving from below the dirt at the front of the cage. They had finally molted and had quickly abandoned the mother; they had been first instars two days earlier. Total time from eggs to second instar was approximately eighty days. Two thorough searches uncovered a total of 32 second instars. Each was moved to a 5.5oz cup with one thumbtack hole in the lid and 1/4" of wet coconut fiber. Three young pillbugs (about 2mm) and six to ten fruit flies, *D. melanogaster*, were placed in each cup.

FEATHERTAIL CENTIPEDE WITH PROTONYMPHS

FEATHERTAIL CENTIPEDE WITH FIRST INSTARS

By the next day, most had eaten every fruit fly. It was amazing how many fruit flies they could go through and most ate more than fifty in the first week. Two died on day seven, so another pinhole was added to the lids in case the ventilation wasn't adequate for the constant feeding.

On day 23 the first molted to third instar. *Alipes* at this stage are close to two inches, including the tail, which is bigger than third instar for some of the biggest centipedes. At this point it seemed they were too big for fruit flies, and their food was switched to small lobster roaches, *Nauphoeta cinerea*, and firefly roaches, *Schultesia lampyridiformis.* Two to three roaches were added a week, but only fireflies after the first few feedings since the lobsters weren't being eaten well.

It took two months for a few to finally molt to fourth instar. A third looked sickly and thin and the skinniest one died. The runts were offered some fruit flies, and it seemed apparent the fruit flies shouldn't have been stopped before they would no longer accept them. One or two lobster roaches that still hadn't been eaten were removed from half the cages. In two weeks every single one molted to fourth instar. Previously, it seemed the fruit flies annoyed the little centipedes, because the pedes went running around the cage swinging their tails wildly whenever a group of fruit flies was dumped in. After seeing how helpful resuming the flies was, this reaction translated differently, like puppies wagging their tails in excitement at feeding time. The first two molted to fifth instar after another three weeks (seven weeks since the first fourth instar).

The female began to eat as soon as the young were gone, but after a few months her appetite decreased markedly and it seemed she could be constructing a new hollow under the driftwood. Was she getting ready to lay a new clutch or was she dying? 7 months, 10 days, after the first eggs were laid, she was holding a second batch! She hadn't molted in two years. Some centipedes molt after maturity, but *Alipes* might not be one of them. Forty days later they changed into protonymphs.

HOUSE CENTIPEDE
SCUTIGERA COLEOPTRATA

Display: 1/2- to 5-gallon cage with damp substrate and limited ventilation. One specimen per cage.

Background: Members of the Old World order Scutigeromorpha are oddities. They are unable to burrow like other centipedes, and have an adapted respiratory system with spiracles on the top. (Spiracles are located on the sides of the body in all other centipede orders.) The blood aids in transporting oxygen. Their eyes look like "bug eyes" since they are made of numerous ocelli and resemble the compound eyes of insects. Even their behavior is different, as their calculated actions contrast the blind thrashing movements of other centipedes.

Scutigeromorpha centipedes are masters of vertical surfaces. In the wild, they have been found living on tree trunks and cave walls. In homes they are commonly encountered on walls. If chased they can jump from the wall and run across the floor at top speed. Young centipedes more readily parachute than adults, and seem to float as they slowly descend to the floor, due to the minimal weight.

Scutigerids are impressive creatures and seem to be made out of legs. Most people are familiar with the common *Scutigera coleoptrata* or house centipede. Adults of this species seldom have a trunk length over an inch, but the leg structure makes them seem much larger. At least one giant tropical Indian species can reach a trunk length of four inches. A few specimens of a giant 2-3″ species from Malaysia showed up in the U.S. in 2009. Despite some success hatching eggs and growing immatures, it seems none made it to maturity.

Scutigera coleoptrata: For most people, this species is #1 on the list of inverts found in the home that sends chills running down their spine. Like house spiders, it is beneficial as it eats insects in the home, but unlike spiders it doesn't leave messy cobwebs and doesn't bite people. Like all centipedes, it is venomous and uses maxillipeds to inject venom into its prey. *Scutigera coleoptrata* is unable to bite through normal human skin, though larger tropical species might be more powerful.

The creature known as the house centipede is the only centipede that's been able to follow man nearly everywhere across the globe. Though a tropical species, it's found in buildings and homes even in the coldest areas, where it ventures outside during the warm parts of the year. In temperate climates, it hibernates during the winter, even indoors. (Basements are much cooler in winter and food is lacking.) But, it will remain active and feeding year round in captivity.

Certainly it is a very fast moving creature (at 1.5 ft/sec it is presently the fastest moving centipede recorded), but if it gets hit with relatively mild impact, it curls up and dies. If the impact hits the legs instead of the body, it drops legs in piles. The pile of legs twitches for up to a few minutes. As long as the animal survives initial impact, the legs grow back close to full size at the next molt. It can also drop legs in stressful conditions, and sickly animals can lose all their legs before dying.

House Centipede

It is strange that the house centipede is so strong and delicate at the same time. It must be a tough critter to follow man and exist nearly everywhere, but it's hard to understand how. Unlike various pest arthropods, there's no part of its life cycle where it is capable of surviving temperature or humidity extremes. It never occurs in large numbers. It's rare to find more than one or two, but both sexes are required for reproduction. Without a drought-resistant resting stage or high reproductive rate (females probably don't produce much more than sixty eggs in a lifetime), it's hard to understand how it has been able to spread around the globe.

Development: Even indoors, egg laying seems to only take place once a year in mid-summer.

Females may only produce eggs twice in a lifetime, as a result of short lifespan and seasonal oviposition. Eggs can be deposited on damp paper towel, but a 1/4″ layer of soil seems to be most successful for laying. (Two eggs were laid on paper towel, so it's possible the female laid many more eggs and consumed them.) The tiny white spheres are approximately 1mm in diameter and are deposited one at a time. Some seem to be glued into the dirt, but most are laid freely. The female should be removed from the cage when eggs are found. The hatch rate for recovered eggs has been 100%. Eggs hatch into a mobile immature in approximately one month. I have been unable to determine the presence or lack of a protonymph stage.

The first mobile immature has only eight legs and looks very much like a tiny cricket. The eight-legged creature doesn't eat, and molts after about ten days into a ten-legged creature as it slowly begins to resemble the leggy adult. At this time the immature begins to feed. At each subsequent molt the animal gains four legs. The leg count on one side over the course of the instars is 4, 5, 7, 9, 11, 13, and 15. At 75°F, immatures took ten weeks to reach 15. The young centipedes are a handsome light purple from leg counts 7 to 13, but they fade to blue and eventually tan. Upon reaching 15 (30 legs), the immature is only about 1/4 grown, but it does not continue to add legs with more molts and it will take at least another five molts to reach full size. Molts slow down to once every three to four weeks, but can take much longer if fed minimally. Molting frequency is incredible compared to other centipedes, and even adults molt every six weeks if fed well. The molt takes place on a flat surface and is not prone to difficulties. They remove themselves head first, like other centipedes, but unlike other centipedes they do not eat the shed exoskeleton.

Food: Members of the Scutigeromorpha feed on small live invertebrates and appear to be the only centipedes that are obligatory carnivores. (Other centipedes can feed on vegetable matter and consume their own shed exoskeletons.) *S. coleoptrata* are capable of taking down prey half their size, but prefer prey only a few times the size of their head capsule. Large prey can damage them. For the first few instars, *S. coleoptrata* are too small to eat hatchling crickets or fruit flies. A captive culture of giant springtails (adults are barely half the size of *Drosophila melanogaster* fruit flies) was

HOUSE CENTIPEDE
(TOP) MOLTING
(BOTTOM) POST-MOLT

used as the food source since normal springtails are too small. Nearly all deaths occurred in the first five instars while being fed on springtails, but it is impossible to know if the survival rate is normal for early instars or a result of prey item nutrition. The survival rate from hatching to maturity was 8 out of 23 separated (6 were cannibalized prior to separation). Fruit flies are excitedly eaten and can be used as temporary food or partial diet for large immatures and adults. However, they lack important nutrients and the animals will not molt again until fed acceptable food. Adults can survive a long time on deficient food, but immatures fed only *Drosophila* flies will not live long.

Habitat: Caging for the adults and nymphs can be very simple and inexpensive. Adults do well in a 16oz deli up with 6-10 thumbtack pinholes in the lid. They can't escape as long as the sides are kept free of debris, but a lid is necessary to slow down drying out of the substrate. 1/4" of any soil-like material will make a good substrate. One side can be sprayed and there should always be a damp area. Frequency of spraying varies greatly with temperature, airflow, ventilation and watering volume, but the trick is just to make sure there's both a dry and wet area. Cage decorations are unnecessary. They seem to prefer to hide and climb, but fare just as well on an exposed horizontal plane.

The immature's cage is the same, but 5.5oz cups are used. Immatures and adults are kept under indirect lighting with 16-hour days, though any day length is likely acceptable. The specimens were kept at temperatures between 70°F and 85°F, but they may be capable of surviving well outside that range.

Although cage cleaning is generally unnecessary, the centipedes still need to be moved to change cage size and for mating purposes. *S. coleoptrata* is unable to bite, but direct handling is a bad idea because the animal readily throws legs and dies from a light blow. To avoid injury, place the lips of the containers inside each other and gently coax the animal to walk over.

Gender and Mating: Gender determination is obvious, but I was unaware of the difference till recently, when I had both a male and female, and the female laid eggs. Size is not an indicator, as the largest and smallest adult specimens I've seen were both males. From above, the male's terminal tergite (last segment) points down. The female's terminal segment is not as sloped and is followed by a small but notable ovipositor. It is possible to easily determine gender when immatures reach 2/3 full size, but not before then. In captivity the only way to determine adulthood is when subsequent molts result in no growth. "Wild" specimen maturity is easiest to determine by time of year, due to seasonal reproduction.

The first attempted mating took place in a ten-gallon with some wood and logs for decoration. Two adults were placed in the cage, but I couldn't determine gender at that time. They recoiled on brief encounters and seemed to show no aggression, but on day sixteen one was missing in action.

Later matings with sexed adults took place in 16oz deli cups and they were left to spend just one evening together. Upon introduction, the male usually backs to the female and tries to guide and direct her using his back legs. The spermatophore is large (~3mm), and seems to be accompanied by little, if any, webbing.

Remarks: Husbandry for *S. coleoptrata* has proven straightforward and simple. Gender determination, mating, housing, development and feeding are likely very similar with large Scutigerids. 2nd instars of the giant species should be able to eat hatchling crickets or newborn firefly roaches, which could lead to a higher initial survival rate.

AMBLYPYGIDS

Sidespotted Whipspider
Phrynus marginemaculatus

Display: 2- to 10-gallon cage with damp substrate and limited ventilation. Numerous specimens can be kept together for six months, but losses add up over longer periods. Rough surfaces such as cork bark laid against the front of the cage will keep them in view, but care must be taken as adults are capable of hiding under very small objects including clumps of substrate.

Background: This small, handsome species is found across peninsular Florida as well as a number of Caribbean Islands. It is specially adapted to its habitat and is reported to survive a full day submerged in seawater (not a normal amblypygid ability). In nature it may be found sheltered under old trees and stumps. Specimens have been found taking shelter in water meter boxes and wandering haplessly into garages.

The unique behavior of whipspiders is endearing to those able to observe it. Specimens carefully unfold elongate, antenniform forelegs to probe their surroundings. The slow, deliberate movements when hunting, courting or just exploring, are fascinating. When startled they can fold up the forelegs and run sideways at breakneck speed.

Sexual maturity is reached at approximately 1/2″ in body length, but adults grow up to 3/4″. However, Caribbean specimens tend to grow larger than those available. Even large, older specimens from Florida rarely exceed 1/2″. As the name suggests, mature specimens are usually dark brown, almost black, with light spots (maculations) on the side margins of the prosoma. 2nd instars have orange pedipalps, dappled markings and a green abdomen, but this color gradually changes over to adult color with each molt.

This whipspider originally was described as *Tarantula marginemaculata,* but that genus name has been popularly used for many Mygalomorph spiders and sometimes for Lycosid wolf spiders. To avoid confusion, the original genus name was suppressed and the next oldest generic description used (*Phrynus*). In order to match the gender of the new genus, it is called *P. marginemaculatus* instead of *P. marginemaculata.*

Sidespotted Whipspiders

Housing: Similar-sized individuals can be kept in community cages. Small specimens hide constantly if kept with larger animals. A 5-gallon aquarium readily houses six to eight adults. More can be added, but problems will eventually arise from stress and territorial disputes. Cannibalism is less likely, but can occur even if specimens are the same size, well-fed and are given sufficient space. An adult will sometimes eat part of another that has recently molted (always resulting in death). If a freshly molted animal is noticed, it should be placed in another cage for a week to allow the exoskeleton to harden.

Sidespotted Whipspider with Eggs

One or two inches of damp substrate is needed to help maintain humidity if ambient humidity is not high. Pieces of flat cork bark or wood leaned up against the sides, or other pieces, should be provided for necessary shelter and molting surfaces. Only one molting/hunting surface is required if kept individually. They molt hanging upside-down and become hopelessly deformed without good molting surfaces. A temperature around 75°F works well.

Food: Tiny prey is ignored, while overly large prey initiates a flight response. Immatures will accept fruit flies and hatchling or early instar crickets. Adults feed well on 1/4″ crickets, as well as nymphs of egg-laying cockroaches such as those in the genera *Eurycotis* and *Blatta*. Burrowing roaches are almost never taken and should not be offered. If crickets are used, it's important to remove uneaten prey within a few days because an imminent molt will result in the soft amblypygid being chewed on and killed by the cricket. They can eat a few prey items each day, but sometimes refuse food for many weeks.

Reproduction: Adults of this species are not easy to sex. Males have slightly larger pedipalps, but it's not easy to quantify. The structure of the genital operculum (plate covering the bottom of the abdomen near the point of attachment to the prosoma) is barely different, whereas in *Damon* and the closely related *Phrynus barbadensis*, the greatly differing structure makes sexing rapid and definite.

Courtship leads to very little physical contact. The male strokes the female with his antenniform legs, which results in hours of back and forth touching if she's receptive. Only in the presence of a sexually mature female does the male glue a spermatophore to a vertical surface, which she can take up under her genital operculum. He may try mating again a few days later, but she can only take up one spermatophore and extras will be left untouched. The white, 1/8″ tall spermatophore has a trunk with two structures near the apex.

Sidespotted Whipspider Eggs
Turn Brown After a Day or Two

A few months after mating the female extrudes the eggs inside a gelatinous, clear sac that hardens over the next few days. Eggsacs are not produced by unmated females. The pastel green color of the eggs is visible on the first day or two, but the egg sac membrane darkens and the eggs appear brown for the rest of the gestation. Unlike the popular *Damon diadema*, the egg sac does not have a flat bottom, and the side view isn't blocked by pleural folds.

Ova hatch after four months into white and green 1st instars that climb onto the mother's back and cling tightly. After nine or ten days, they molt into second instars that are ready to disperse and will soon hunt. Immatures can be kept together as long as they are within one molt of each other.

SIDESPOTTED WHIPSPIDER WITH FIRST INSTARS

Remarks: Phrynus are easy to house and more accepting of each other than most larger amblypygids. Keep in mind that despite gregarious tendencies, collection reports consist almost entirely of individuals or pairs. Although this species had been difficult to acquire for a number of years, it has been consistently available in recent years. Don't miss the opportunity to display this spectacular, oddly-behaved and non-venomous arachnid.

GIANT TANZANIAN WHIPSPIDER
DAMON DIADEMA

Display: 5- to 30-gallon cage with 5-10% of the lid ventilated. One adult male and a number of females can be displayed together.

Background: These bizarre arachnids are gigantic, long-lived and surprisingly harmless. *Damon diadema* from Tanzania, Africa, is one of the world's largest Amblypigid species. The legspan is a maximum half a foot across, and the 'whipspan' a foot. Mature *Damon* vary in size, and a large adult can be more than twice the size of a small one. Size is gained in post-maturity molts, so the largest individuals are often the oldest. Adults live multiple years, but are more likely to obtain 5 years than the hypothetical 25-year mark suggested in some writings. The huge pedipalps tear apart crickets, but seldom scratch human skin.

Damon is an archaic spelling of the word "demon." The species name comes from the word "diadem"—the king's crown—so the scientific name could be translated as "the demon king."

These are arachnids from the order Amblypigi. Some of the animals that have been coming into the U.S. market over the last five years from Tanzania have been mislabeled as *D. variegatus*. The two species look very similar, but *D. variegatus* are not from Tanzania, have pleural folds covering most of the egg mass, and have a minor spine difference on the pedipalps.

Husbandry: Amblypigids do not molt off the bottom of a cage. If they survive a bad molt, they'll lose a whip or two and tend to grow slowly and can have a hard time feeding. (Whips grow back at half size or better in a single molt, but can still be undersized after many molts.) Flat pieces of wood or bark should be placed at a high angle to provide adequate molting surfaces. Humidity and feeding are also important, but trivial by comparison.

They eat only active, high-profile foods like crickets, flies (flightless) and certain roaches like *Periplaneta*. Amblypigids are more likely to die of starvation than feed on food that hides, including most feeder roaches, mealworms, isopods, etc. Their appetite varies greatly—sometimes they'll eat twice a day and sometimes they won't eat for weeks.

They do well together, but only for the first few molts, and then when mature or nearly mature. (They don't get along well in the "teenage

Giant Tanzanian Whipspider with Babies

Giant Tanzanian Whipspider, 4th Generation Adults

years," and should be separated.) In groups they can be fed well without the need for close monitoring, and kept in a large cage with many molting surfaces.

Still, Amblypigids do cannibalize. I lost the first one after nine months and lost two more over the next two years. For a long time I thought the deaths were exclusively due to hungry crickets, combined with bad timing. They had only been partly consumed, which is consistent with crickets. Much of the available literature claimed they don't bother each other and I witnessed a few molts where they seemed to watch over each other. However, I chanced upon the last one to die. It had just molted—it was green—and half of it was sticking out of its sibling's mouth. I had to wonder whether any of the previous deaths were caused by crickets.

They don't grow very fast. With super close attention and constant feeding, it's possible for young *Damon* to molt every few months. My older nymphs molt every 6-10 months and adults every 12-18.

A few months after mating, the female was carrying a mass of eggs. She continued to eat a cricket a week. Female tailless whipscorpions continue to feed while holding eggs. A female might survive incubation without feeding (as suggested by some authors) but at best it might not kill her. Eggs take a long time to hatch and during gestation are internally nourished. The female can be kept with other adults until just before hatchlings molt and disperse. Hungry adults would most likely eat the little guys.

It was a long three months of gestation before I was certain the eggs were fertile. I could see notable development of the eggs through the semitransparent membrane. It was another month before they hatched.

The membrane began to swell forty-eight hours before hatching, and began to crack eight hours before hatching. Three of the hatchlings didn't make it to her back and hung out together in a depression in the dirt. The female doesn't feed hatchlings or provide any visible care, and

FRESHLY MOLTED GIANT TANZANIAN WHIPSPIDER

the humidity was high throughout the cage so I left them. They seemed fine, but on the morning of day three, two were dead. I carefully coaxed the survivor onto the mother's back.

On day eleven, 34 young molted out and dispersed. I had accidentally allowed the wet substrate to become barely damp and three lost a leg or two in the molt. They were partly stuck in the old exoskeleton and had to be cut free.

The young were moved to a 1.5 gallon bucket with an inch of wet substrate and three 5″ x 4″ pieces of egg carton. The center of the lid was cut out and covered with microscreen. They were fed flightless *Drosophila hydei* and pinhead crickets until their abdomens swelled like little green balloons. It was surprising how much they ate.

Thirty-nine days following the dispersal molt, the first baby *Damon* underwent its second molt. One or two continued to molt every few days at a regular pace until the second to last one molted on day seventy-four. The three damaged immatures were among the last ten to molt. One straggler didn't have any damage or visible reason for molting late and finally molted on day

Giant Tanzanian Whipspider Pair, Male is on the Left

Giant Tanzanian Whipspider with Eggs Starting to Hatch

119. The first to molt for the third time also molted on day 119. Three weeks after this molt, they began to cannibalize and were separated. On day 187 they began to molt for the fourth time.

Rearing containers for young consist of 16oz deli cups. (32oz may be better for a sub-adult or adult.) A piece of metal window screen is glued to the underside of the lid to provide a molting surface. A section of plastic (used for yarn crafts and available at any craft store) is cut the length necessary to reach from the bottom corner to the top, at an angle. This plastic is important and used as a ladder since the amblypigid can't climb smooth sides of a container. One 1-2″ square is cut into the side and a piece of microscreen or cloth is glued over the hole. Larger screens can be used but increase problems with drying out. Once the cage is built, a 1/4″ layer of coco fiber or soil is placed on the bottom for moisture retention.

Adults: The female molted 189 days after the dispersal molt of her young, and increased more than 30% in size. The male molted 15 months after arrival (just before his children hatched). I was impressed to see he had gained 20% in size. While still green, he fell four feet to the cement and died the next day.

I was able to acquire another adult male to be placed with the adult female. They coexisted "happily" in a small cage. He was much smaller than other mature males I'd had, but he did make a few spermatophores. Months later a new female was placed with the pair. She seemed to be an outcast, so I only left her in for a few days at a time. I hadn't considered she would be killed. On day three of the fourth attempt, she was on the substrate and was very sluggish. She had no missing parts but on close inspection her prosoma had two half-inch lacerations—evidenced by dried blood. She was isolated immediately, but died after a few days.

Remarks: Important points:

1. Don't handle recently molted animals.
2. High humidity is always helpful, but is paramount during gestation and the first few molts.
3. Females should be fed regularly when holding eggs, and can be separated any time before young molt and disperse.
4. Older young cannibalize—separate at the third molt.
5. Provide rough, high molting surfaces. The distance should be about twice the length of the walking legs. Some hobbyists prefer much higher molting surfaces citing the long front legs. However, the long front legs are carefully extracted from the exoskeleton by the pedipalps during the molt, so extra distance is unused.

GIANT AMAZONIAN WHIPSPIDER
HETEROPHRYNUS BATESII

Display: 5- to 30-gallon aquarium with solid glass lid. One adult male and a number of females can be displayed together.

Background: Although the 12″ legspan of the male *Theraphosa blondi* tarantula is often touted as the greatest of the arachnids, it's not technically true.

HETEROPHRYNUS, 2ND INSTAR

Heterophrynus batesii, Adult Male

The whip-like antenniform front legs of a few amblypigids can be as much as 24″ in span. *Heterphrynus batesii* is one of these. This species has very long legs it sits up on, and the attitude resembles a gigantic opilione rather than the crab-like profile of the amblypygid species described above.

The walking legs are long, but the male palps are insane. It probably holds the record for arachnid pedipalp span, with expanded palps exceeding 8″. Members of this genus are known to hunt on tree trunks in the rainforest.

This species of *Heterophrynus* has been kept in captivity in the U.S. for decades, but is still extremely difficult to find compated to the above *Damon* and *Phrynus*. Other, smaller *Heterophrynus*, including a small, short-palped species from French Guiana, have made their way overseas, but few were imported and even fewer survived shipping.

Husbandry and Life Cycle: Females have shorter palps than males, but otherwise aren't easy to sex. Mating involves hours of antenniform leg stroking, and ends in the uptake of a diamond-shaped spermatophore. A few months later, the eggs are extruded in a membrane. The hanging mass and centrally upcurved opisthosoma is very noticeable. No pleural folds protect the eggs and they do not hang upside down, as the membrane hardens to flatten the membrane. After about three months, the helpless first instars hatch and hang onto the mother's abdomen until the first molt takes place in ten days.

Young specimens can be kept together. Adult males will likely fight and kill each other, but seem unaggressive towards other males compared to *Damon*. They need relatively tall molting surfaces due to the long legs and extremely long whips. Slanted pieces of cork bark twice the height of the walking legspan work well. Aquariums with solid glass lids are suggested to maintain humidity, as the small gap between the lid and aquarium rim will provide enough ventilation.

Burrowing and hiding food is rarely caught, so house crickets are usually the primary food offered in captivity. However, crickets will kill molting *Heterophrynus* immediately, and should never be used without extreme caution. Firefly cockroaches, *Schultesia lampyridiformis*, are the recommended food, though caging must be sealed appropriately. These cockroaches never harm the *Heterophrynus*, and are consumed with gusto. Other non-burrowing genera like *Blatta*, *Eurycotis*, or *Dorylaea* might also be useful feeders. Prey items should be offered about once a week.

WHIPSCORPIONS

GIANT VINEGAROON
MASTIGOPROCTUS GIGANTEUS

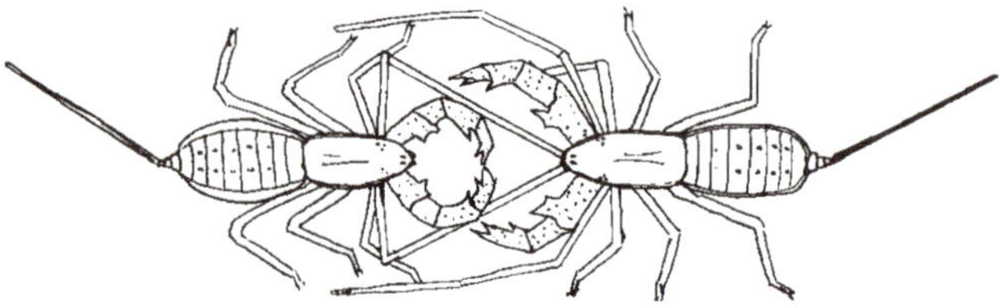

VINEGAROONS MATING

Display: 2- to 5-gallon cage with good ventilation and a water dish. Specimens cannot be kept together. Substrate should be shallow, or they will hide from public view.

Background: This is one of the more spectacular invertebrates for use in displays. It has a frightening appearance, but is harmless and can be handled with minimal concern. The primary defense, hinted at by the common name, is an ability to project a defensive fluid that contains acetic acid and therefore smells like vinegar. While it sounds no more harmful than vinegar chips, if the spray gets in the eyes, it would not be comfortable.

Mastigoproctus giganteus are readily available as adults or immatures. Captive-hatched young

IMMATURE VINAGEROON

are also commonly available. Some much smaller tropical specimens like *Thelyphonus* have been available. These small species are similar, but don't require much ventilation and some can be kept communally.

Vinegaroon Pair

Husbandry: Mastigoproctus are hardy feeders and consume nearly any invertebrate offered to them, often within minutes of introduction. Large specimens can be fed pinkie mice, but there is no need for vertebrate prey. Specimens seasonally forgo food for months in nature, but prey should be offered once or twice a week in captivity. A water dish or damp substrate should always be available to supply moisture. A

Two Female Vinegaroons Fighting

cover with multiple vents or a full screen lid is suggested.

Adults normally live one to five years in captivity. Immature specimens are very slow growing. Although they only molt four times, it takes no less than one year per molt, and immatures can go a few years between molts. For longevity, it's best to acquire a large immature. Captive-hatched specimens have been common in recent years.

Vinegaroon with Eggs

Adults can be mated year round in captivity, but normally mate in later summer. The mature male has larger, triangular-shaped pedipalps, while females are smaller and rounded. Courtship can take half a day, as the male dances with and drags the female around by her antenniform legs. Mating does not look aggressive and cannot be mistaken for fighting. If two same-sexed or uninterested specimens are placed together, one usually eats off the other's tail (flagellum) and eventually kills the other. Pairs usually can be kept together for a few days without aggression.

SPIDERS

Golden Silk Spiders
Nephila species

Display: 10-gallon minimum cage with damp substrate and limited ventilation, but stagnant conditions can cause large spiders to mold and die. Watch for mold growth in the habitat. Usually one female and a male are housed in a cage, unless it is very large. Adult females may be "free-ranged" in greenhouses.

My Introduction: In 1994 I stopped at a 24-hour-breakfast restaurant off the highway, halfway down the eastern coast of Georgia. After eating I took a walk around back. Bird grasshoppers and cone-headed katydids jumped and flew as I walked. As I tried to grab one of the grasshoppers, a two-inch katydid flew up and was stopped dead by a nearby web. The web didn't even shake! An improbably large spider scurried down and wrapped it up.

I own many books with information and pictures of *Nephila clavipes*, but I had never noticed it. In real life, this spider forced my attention. I stepped into the tiny cypress and palmetto swamp behind the restaurant and entered a world I had previously only known from science fiction movies. Every six or eight feet, monstrous webs were anchored between tree trunks six to eight feet apart. Some webs were high up in the trees, but I could still see them with their improbable owner sitting patiently in the middle. The web size and strength was eerie. It almost felt like I could get caught in one.

These weren't the orb weaver webs I was used to. I'm usually a few feet past webs before I know it. I only know I hit a web because of the annoying tickle of the invisible silk strand on my skin. *Nephila* webs are different. They are visible at almost any angle, and I feel the pull of the anchor lines as soon as I walk into it.

Nephila inaurata, Pair

Background: Golden silk spiders (AAS common name "golden silk orbweavers") are massive web-building spiders found across the world's tropics. They are the largest members of the family Tetragnathidae, the long-jawed orbweavers. Even the smaller *Nephila* species have four-inch legspans. They build improbably oversized webs commonly slung between tree trunks rather than twigs. Their name comes from the thick, golden yellow silk strands used to construct the webs.

Nephila possess huge chelicerae and a fierce appearance. Within their webs they are ferocious predators capable of rapidly immobilizing a two-inch cicada in seconds. Without their webs, they

NEPHILA CLAVIPES

are like a fish out of water and won't tackle a gnat.

Despite Madagascar's claim to many of the gaudiest inverts, its native *N. inaurata* is no bigger than (and plainer-looking than) *N. clavipes*, which hails from the southern United States—south all the way to Brazil. These two *Nephila* species are the only *Nephila* that have been kept with any regularity in the U.S. I've successfully reared groups of both species in the manner described here. *Nephila plumipes*, *N. senegalensis* and an African *Nephila* sp. are similar species cultured in Europe.

Husbandry: Eggsacs are kept humid and hatch after two months. They survive a little neglect, but more than a few days of dryness will kill the eggs. Spiderlings remain inside for months, and begin to leave a few weeks after the second molt. They start out in communal webs but immediately set about building miniature orb webs if separated. Their legspan is less than an 1/8", but webs are a few inches across. About fifty are placed per 10-gallon tank. (A single eggsac usually hatches 250-300 spiderlings.) The bottom is filled with an inch of damp soil, apple halves and a ton of fruit flies—flightless varieties work better for web hunters. The lid is screen, but covered over with plastic wrap to retain both humidity and fruit flies. (On the grounds of a small zoo in the Bahamas, in 1998, I learned nature employs the same method I had been using. There were hundreds of young *Nephila* sp. in a small area below a fruit tree surrounding piles of rotting fruit and clouds of fruit flies.)

Cannibalism is lessened with constant feeding or by separating spiderlings after a few molts into 32oz deli-cups. After six to eight weeks in a 10-gallon, there will typically be fifteen or so surviving spiders with legspans between 1/2 -1 1/2". At this time they need to be separated, or in another month cannibalism will reduce the number to three or four. Males are smaller and ought to be eaten first, but usually make up 2/3 of the survivors. (It is possible that 2/3 or 3/4 of the eggs hatch males. In the wild, each female's web contains 2-4 males.)

Place 3-5 spiders in a 10-gallon or large pretzel jar, and remove extra females as they become

NEPHILA EATING CICADA

obvious. A few males can be kept with each female, but there should be only one female. Even at an inch, it can be difficult to tell male from female if the male palps aren't yet developed.

Adult males come in two forms. One has a 1/2" legspan, with smooth legs, and holds its body like a crab spider. The other form has a 1" legspan, and sparsely spined legs (*N. inaurata*) or sparsely hirsute legs (*N. clavipes*). The larger form resembles an immature female, except palps are easy to see in the last few instars. In captivity the forms seem to occur in roughly equal numbers. However, in nature the smaller form male of *N. clavipes* isn't common.

The two-inch female constructs a web large enough to handle small crickets and large moths. If prey struggles free, try a different size or different type of animal, or offer the spider a better place to build her web. The more space she has, the stronger and more complex the web. Do not throw in prey likely to tear up the web. She does not like her web torn apart. If it's mangled constantly, she will stop rebuilding and starve.

Spiderlings are started on *Drosophila melanogaster* and switched to the larger *D. hydei* when around 1/2". They accept the larger fruit flies till obtaining a1.5-2" legspan. By this size, hand-feeding crickets will work but is tricky because it is important not to damage the webs. Fly larvae can be purchased, kept warm and placed in the cage as puparium. Large meal moth cultures must be started before the *Nephila* eggs are laid, since the moths take a long time to reach adulthood. Adult green banana roaches, *Panchlora nivea*, are an excellent food as long as the cage is secure. In summer, plenty of flying food can be collected around the porch light.

DISPERSING SPIDERLINGS

When the female reaches adulthood, a 10-gallon aquarium or giant pretzel jar are small cages. Captive-raised females have no problem making webs in such cages, while wild females may refuse. In a small cage she may seem stressed, construct a sorry little web, and remake it every few days. Observe her behavior and move to a different size or shape cage if necessary. When her legspan is a few inches and she's about two molts from maturity, I set her loose in a corner or between furniture to let her make a web in the room. (If released at a smaller size, survival is zero.) She remakes the orb once every week or so but reuses anchor lines and stays in the same spot.

Make sure to use a lid with a door, otherwise the web will be warped and ruined every time the lid is lifted for feeding. If a tall plastic jar is used, a 3/4" hole—feeding port—can be drilled in the lid. A piece of sponge is cut and used as a stopper.

Feed 2-3 times a week and lightly mist daily, but not at the same time. Excessive spraying or major disturbances can cause her to tear down the web.

Do not introduce males before her web is established. Males can be introduced at the furthest edges of the web. Males mature and die more quickly than females. Two or three can be kept with one female, even when she's immature, so they'll be around as soon as she's ready. They hang out on her web and eat tiny catches she ignores. (Even the larger form males eat prey the size of fruit flies, while a large female has no interest in food smaller than a housefly.) The males only get eaten if she is starving. Males from the same eggsac can die before she's old enough to mate. This is a result of less tolerance to imperfect conditions and a shorter lifespan. Larger form males are hardier, but more likely to get eaten.

Nephila require high moisture. Excessive moisture doesn't harm spiderlings, while dryness rapidly kills them. They should be housed in a covered (not airtight) cage with wet substrate. Misting is not practiced in a good cage, because the high humidity will not allow the droplets to dry out. The tiny adult males are also prone to desiccation. Adult females have a lower surface area to volume and can survive dryness that would rapidly kill smaller spiders. Still, females shouldn't be "free ranged" if ambient humidity is less than 60% and if daily misting isn't possible. The "free ranged" female will need to be misted daily if the humidity is below 80%. She might survive weeks without misting but can die at the next molt, or permanently discontinue feeding.

Female *Nephila* produce 3-5 eggsacs the size of small marbles, surrounded by much larger semitransparent webbing. One sac is laid every two to three weeks until the female dies. Each is webbed to a flat, vertical surface just beyond the periphery of the web. Eggsacs will not survive the low humidity of a home and should be suspended over a few drops of water or damp substrate, in a minimally vented cage. Months later, the magnificent giant females are gone but uncountable tiny legs rumble inside the silken spheres.

River Orb Weaver

Display: 2-gallon or larger cage with damp substrate.

Background: This handsome spider is usually found near rivers and other groundwater across much of the United States, where it feeds on midges. Webs are often coated with the overabundant prey in late summer when females use the copious nutrients to form up to a dozen eggsacs the size of a small marble. It's one of our larger orb weavers based on mass, bigger than our common *Aranea* species. A full-size female has a 1 1/3" legspan and a rotund, circular abdomen up to 1/2" across. Males grow nearly as large, but have smaller abdomens and easily seen palpal bulbs. It is nocturnal in nature, as it feeds on midges and flies that emerge in mass from rivers at night. Specimens tend to come out to feed when food is offered during the day in captivity. They are found in large numbers and anchor lines often cross, though they are cannibalistic. It's interesting that in the "wild" orbs are nearly always found close to artificial light sources. This spider is probably *Larinioides cornutus,* a species

RIVER ORB WEAVER

commonly found throughout North America and Europe, and which occurs globally.

The orb web of an adult is the classic spider orb, though it's large at two to three feet across, usually larger than *Argiope* orbs. The anchor lines are commonly only one to two feet further, though some webs I've seen in trees overhanging a river (and near lights) have anchor lines to ten feet.

Husbandry: This is an aggressive species and the only orb weaver I have observed that feeds readily without a web. On the other hand, despite large natural webs, an adult female can be content forming a strong orb inside a tiny 32oz deli-cup. Flying food is best, but cockroaches can be thrown in the webs. Hatchlings are easily reared up on flightless fruit flies in deli-cups, with a dozen thumbtack holes in the lid and a thin layer of damp substrate on the bottom. They are similar in care and habits to *Nephila* and *Argiope*, but are a lot more accepting of a wide range of conditions.

Extreme dryness is not tolerated, but most caging is acceptable. This is a great species for less advanced keepers, as it's much hardier and longer lived than most orb weavers and males live almost as long as females. Adults and immatures hibernate through winter, and with hibernation they appear to live a few years. However, in captivity the lifespan is usually six to nine months.

THE REGAL JUMPING SPIDER
PHIDIPPUS REGIUS

Display: 1- to 2-gallon cage with damp substrate and good ventilation. Specimens should be kept singly.

Background: Phidippus regius is an amazing looking species found commonly across peninsular Florida. It is uncommonly encountered in other parts of the southeastern USA. It also occurs in the Bahamas. *Phidippus regius* resembles the widespread eastern *P. audax*, but is much larger, and the male's sexual characteristics are more exaggerated. *Phidippus* is a large genus with numerous species found across North America.

P. regius is huge for a jumping spider and among the largest of its family. Gravid females can grow to a whopping 22mm in body length.

The Regal's Family: While various authors agree that the family Salticidae (jumping spiders) is the largest of the spiders, the estimated number of known species seems to be different in every popular literature book: 4,000, 4,400, 5,000, etc. Of course some named species are likely synonyms and countless more await discovery. Two tiny species from Mount Everest hold the world record altitude for spiders, at 22,000 feet. Overall they are small spiders—the body length of the largest of the world's salticids is under an inch. Members of this family grow and live quickly. The usual life cycle is one year from egg to egg.

Jumping spiders are a favorite of hikers and nature enthusiasts, due to their energetic motion and perceived personality. They are among the few types of day-active spiders, and are outfitted with a pair of large, forward-facing eyes (with additional pairs to the outside and on the dorsal surface of the prosoma). Humans commonly associate large eyes with intelligence and even cuteness. Forward location of the eyes and lack of eyelids gives the impression of intense interest and "intelligence" to the casual observer. Locomotion involves tiny jumps and skittish movement, and appears somewhat comical.

Female Regal Jumping Spider

Regal Jumping Spiders, Immature

Although not a stranger to vendor lists, the small size, short lifespan, specific requirements and diligent care required to successfully rear jumping spiders keeps popularity at bay.

Reproduction and Husbandry: The *P. regius* male and female are incredibly different in color and form. With oversized, furry, front "arms" and a large, oval "head," the male is reminiscent of a miniature gorilla as opposed to the classic jumping spider appearance of the female. The male's opisthosoma (abdomen) is much smaller, but the larger prosoma and legs afford him a larger, more menacing appearance (if you're scared of really tiny gorillas). He is black overall, with white markings and oversized, metallic-green chelicerae. Her chelicerae are an amazing purple, but, unlike his, are covered by the palps and often go unseen. Females come in a variety of shades and colors, from straw to dark brown to brick red.

Visual cues from the male keep her from eating him during mating, which is likely the reason for the consistent coloration of the male gender, versus the variable and cryptic female gender. Her recognition is more important than general predator avoidance.

After mating, the female forms a web larger than herself and places a large, rounded disc of eggs in the center. The few eggsacs I've seen contained fewer than 150 eggs, but they supposedly can exceed that number a few times over. The mother feeds minimally and guards the sac, but the eggs hatch just as well if she is removed. With or without the mother, high humidity must be maintained for survival and development. Within about forty days, the spiderlings have hatched, molted twice and are ready to leave in search of food. The female may produce multiple sacs, but generally makes one and dies less than a month after the spiderlings have set off on their own.

MALE REGAL JUMPING SPIDER

Initially, I wrongly assumed a terrestrial rather than arboreal feeding habit. The dispersing young were placed in a 16oz deli-cup with a dozen pinholes in the lid, a shallow layer of damp potting soil and a number of 1-1.5mm "giant" springtails. (Fruit flies were not attempted because even the smaller *Drosophila melanogaster* are twice the size of the spiderlings and do not provide adequate nutrition for some terrestrial arachnids.) The 2nd instars immediately constructed a communal hammock of webbing around the inside diameter of the lid. I thought they'd come down to eat when ready, though feeding was not observed. It was uncertain if they were old enough to eat or already feeding. After a few weeks, the spiders began to shrivel up from starvation. I added some fruit flies and within a few hours the survivor's abdomens expanded like balloons from feeding (confirming the previous lack of feeding). From the small abdomens, it was certain that not one had previously eaten a springtail or cannibalized a sibling.

Fruit flies appear to be an adequate diet for *P. regius* for the first six or more instars. An alternative is to use 1st instar roaches of a small glass-climbing species such as *Schultesia lampyridiformis*. Springtails or early instar crickets may be used as long as the cage is designed in such a way as to make the prey wander into the predator. Older spiderlings and adults do well on small crickets.

Each animal can be contained in a two to five ounce plastic cup with a few pinholes and a layer of substrate to maintain moisture. Larger cages are unnecessary and can make feeding more difficult. Decorations are likewise unneeded, and ignored in small caging.

Silk and humidity are integral components for successful molting. Immature spiders build tube webs on the lids of the containers almost like miniature *Avicularia*. The silken hideaways are used as shelter, but the more important function is as a refuge for molting. Nevertheless, if a bunch of running flies are fed before a molt, the spiderling often becomes stressed and leaves its webbing. Molts outside the web result in a few stuck legs. Minor leg damage during molting is rarely life-threatening for other spiders, but relatively minor molting issues result in death for *P. regius*. Silken shelters provide no protection from low humidity, so substrate should be kept damp and ventilation should be restricted, or molts inside the web will result in damage.

Remarks: Regal Jumping spiders are energetic, metallic beauties that can be successfully reared with minimal requirements and available feeders. Spiderlings are very sensitive to low humidity and disturbances. The spectacular adult males are well worth the close attention required for them to successfully reach maturity.

MEXICAN REDKNEE TARANTULA
BRACHYPELMA SMITHI

MEXICAN REDKNEE TARANTULA

Display: 5- to 10-gallon cage for one adult. Shallow, dry substrate, a tiny water dish and good ventilation should be provided. No hides. It is docile and rarely flicks setae. Handling often results in a ruptured abdomen and a dead spider.

I've seen a number of tarantulas in zoo displays where they are given ridiculously large display cages with huge hiding areas, or the display features *Avicularia* or another arboreal with massive web structures. If the tarantula were to die, nobody in the public would ever know. Displaying something nobody can see is the same as turning on a light and putting a bowl over it. Hides do not improve the "emotional state" of tarantulas. Large terrestrial species that sit out in the middle of the cage should be chosen for displays if webbing can't be removed daily. There are dozens of common species that fit the bill and are cared for similarly. *Brachypelma smithi* and the following species are the two most famous and best large terrestrial display tarantulas.

Background: This handsome display animal is often credited with starting the tarantula hobby.

Wild adults were commonly available at pet shops for a few dollars in the 1970s, but due to worries they could be overcollected, the entire genus was added to CITES and wild specimens are not imported. Other colorful members of the genus *Brachypelma* from Mexico such as *B. boehmei, B. emilia* and *B. vagans* are similar and very closely related. (A number of *Brachypelma* have been crossbred in captivity.)

Small spiderlings are available at reasonable prices, but it would take a good five years to get them to maturity, and can take longer for maximum size. Large females command a premium price. Many wild-caught females have lived twenty-five years in captivity. Males are less expensive and only live a year or two after the five years or so required to reach maturity.

Brachypelma smithi is regularly bred in captivity, but it is a difficult species to successfully reproduce. However, one eggsac can contain up to a thousand eggs, so it doesn't take many successful hatches to keep the species available.

Husbandry: This is an arid land species, so limited ventilation and high moisture will kill large specimens. Small spiderlings are more accepting of dry conditions than many species, but should never be kept dry. Until the legspan is a few inches across, ventilation should be limited to prevent desiccation and unsuccessful molts.

Prey items include readily available crickets and mealworms, commonly cultured cockroach species, and pinkie mice. Young spiders can be fed small cockroaches, crickets and adult cowpea weevils. Never use fruit flies solely, as nutritional deficiency causes spiderlings to molt into sickly animals with twisted legs. (Fruit flies are fine as a sole diet for mantids and web spiders, but not for whipscorpions, tarantulas, or centipedes.) Hungry specimens will accept prey larger than themselves, but smaller prey should be chosen. Prey that buries itself like mealworms and some types of cockroaches may evade capture indefinitely in large, planted cages with deep substrate.

GOLIATH BIRDEATER
THERAPHOSA BLONDI

Display: 10- to 20-gallon cage with damp substrate and good ventilation. One spider per cage. This highly-defensive species has huge chelicerae, strong venom and bothersome urticating hair. You may have seen the news clip where an animal presenter holds one in his hand and drops it a short distance. Its abdomen bursts and guts leak out. He pulls a quick save, saying she's laying eggs, which sounds reasonable to anyone who doesn't know they just watched a magnificent specimen die.

Background: Goliath is a descriptive term, though it is likely this "birdeater" doesn't run into many birds on the forest floor. It is the world's largest known arachnid. Only the very similar *Theraphosa apophysis* grows as large. The male has a greater legspan, up to 12", but the 10" female is the larger of the two and her abdomen can be the size of a small orange. The female is a handsome tan suede while the male is brown. *Theraphosa blondi* is a New World species found from Brazil and Venezuela to French Guiana.

Husbandry: Spiderlings are large and somewhat temperamental till they've gone through a few molts. The early stages of this species are notoriously fragile and egg sacs fail to hatch regularly. As individual eggs are large, the sacs contain relatively small numbers, from 80-150.

With reasonable precautions, the female is unlikely to eat the male during mating attempts. He doesn't have tibial hooks to hold her at bay, which is often the case for species with less aggressive females. Eggsacs range from 30-100. The large eggs and early instars are sensitive and easily lost.

Photo Credit: © Louise Efer

GOLIATH BIRDEATER

PSEUDOSCORPIONS

Anvil-Clawed Pseudoscorpion
Microchernes dentatus

Display: Even the largest pseudocorpions are extremely small and difficult to display, but can be useful for up-close study.

Background: The anvil-clawed pseudoscorpion is a handsome species commonly found in northeastern North American forests. It is one of the few easily identified pseudoscorpions, due to the presence of an odd, anvil shaped process on the hand of the male's palpal chelae. Specimens are usually found in rotten logs or leaf litter, where they feed on mites and springtails.

The specimens depicted here are descended from a group of less than a dozen collected in a rotten log in Brunswick, Ohio, in summer 2004. Pseudoscorpions are commonly encountered in rotten logs, but this was an unusual log. The fungus consuming the damp wood had a sugary smell and was highly attractive to mites. After a few weeks the wood was discarded because it had turned into a nearly solid mass of tiny grain mites. With exception of dying fruiting bodies, fungus that consumes rotten wood is inedible for grain mites. I've never seen wood like this before or since. The species of tree the log came from is uncertain, but it was a hardwood, possibly beech, and came from an area containing mostly oak, beech, and maple.

Cohabitation: The anvil-clawed pseudoscorpion and most, if not all, species in this order are gregarious. Despite predatory habits, the adults don't appear to eat immatures even with great size discrepancy and close proximity. This is likely the result of a species-specific odor or tactile uniqueness. It's not the result of an inability to attack and kill other pseudoscorpions. I've only tried placing individuals from different species in the same cage a few times, and even in a relatively "large" 16 oz deli-cup they hunt each other down. Within a few hours one is found dead in the pincers of the other.

Gender and Reproduction: Adult males and females are easy to tell apart. A glance at the pedipalps is all that is necessary. As far as I know this is the only species in the world with this feature. As it is gender specific, the anvil process is probably used by males for sparring over females. There may be another purpose, as males are found in close proximity.

Spermatophore production isn't necessarily accompanied by courtship in pseudoscorpions. The anvil is outfitted with an accessory sensory organ that may be analogous to the cheliceral flagella of the related windscorpions. Males have thinner, heavily-armored abdomens and somewhat thicker pedipalps.

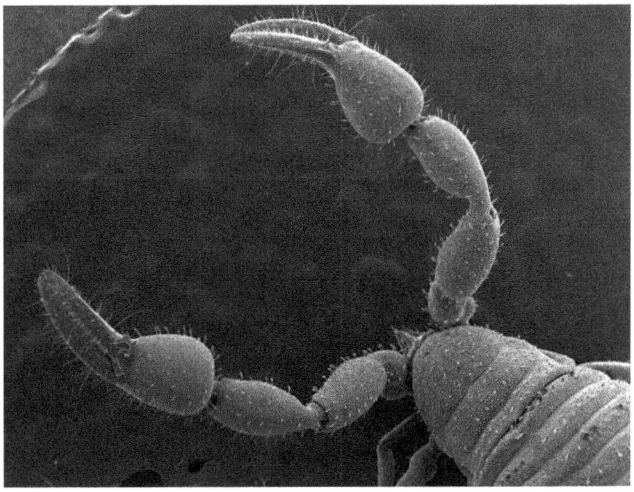

Anvil-Clawed Pseudoscorpion, Female
(Scanning Electron Microscope)

For the first few years reproduction was slow and limited. Fifteen or twenty were reared from young that hatched only in the summer. These took nearly a year to mature. A small amount of prey was offered once a week. Feeding frequency and volume has since increased. Dozens can now be reared annually and broods are not restricted to a time of year.

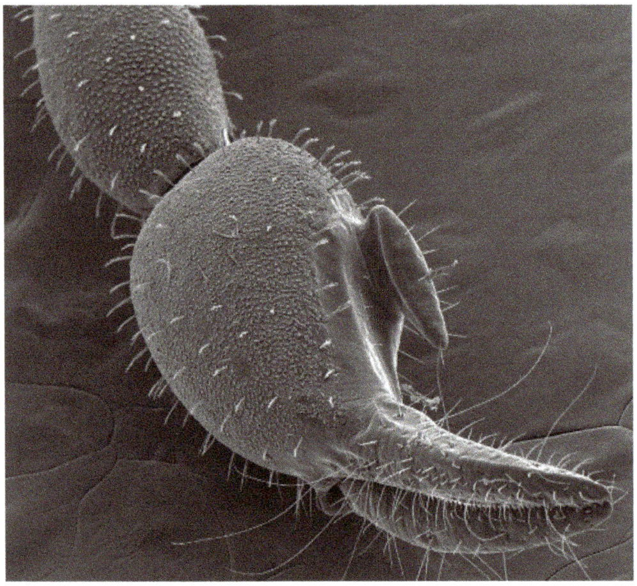

ANVIL-CLAWED PSEUDOSCORPION, MALE'S CLAW (SCANNING ELECTRON MICROSCOPE)

Eggs are attached under the female's abdomen. Initially they are small compared to the abdomen, which remains very portly. When they hatch into postembryos (or the 2nd embryonic instar), they gorge themselves on nutritive fluid from the female's abdomen and triple their size, causing her abdomen to thin out. Since they face inward it appears as though the eggs have grown greatly in size. The female holds first embryonic instars while she resides in a brooding chamber constructed of silk and covered in debris.

Caging: Microchernes dentatus cultures can be permanently housed in a 16 or 32oz deli-cup. If the colony grows to more than fifty adults it can be split, as high density can make it difficult to maintain prey levels, and overcrowding will likely prevent further reproduction. A layer of damp, rotten wood is the best substrate. Large pieces on top provide areas for hunting and forming molting webs. Ventilation holes should be a small number of pinholes, so water does not have to be added to the substrate more than a few times a year. Dampness is needed, but standing water and droplets can drown them. This species is temperate and survives extreme cold, but in captivity it can be kept constantly at 72-77°F. The effect of higher temperatures is unknown.

Prey: Adults and older young eat fruit flies, hatchling crickets and other tiny insects, but this colony has been fed primarily mites and springtails. Invert food overrun with grain mites can be thrown in the cage, or dead roaches with running brown scavenger mites can be shaken over the culture cage. Mites were used as the primary diet originally, but the easily cultured "giant springtail" has become nearly the entire diet, since it's hard to collect enough mites to keep up with the pseudoscorpion culture.

MICROCHERNES IN NESTING CHAMBER

Remarks: Despite having mastered husbandry through the process of developing a large, stable colony, I realize there are many questions about this species I can't answer. The manner in which spermatophores are transferred and the purpose of the anvil are the two most glaring. Social habits and specific gestation times are others. However, the information I've learned so far is more than enough for successful culturing, which is the first step to answering other questions.

WINDSCORPIONS

Display: 1- to 10-gallon cage with screen or vented lid. One animal per cage. Specimens seldom last a month or two.

Background: Members of the arachnid order Solifugae are extremely energetic and ephemeral arachnids. Their unique look and amazing energy levels enthrall the public. On the other hand, short lifespan, difficulty of keeping, and inexplicable mortalities severely limit solifuge displays. Large African species, often *Geleodes*, are available from vendors seasonally.

Eggs and early instars commonly do not develop, resulting in a scarcity of reports on the particulars of hatching and early development. The following describes the early captive development of an *Eremobates* species from the southwestern United States.

Ova and Early Instars: Two females were originally collected the last few days of July 2009 in Rio Rico, southeastern Arizona. The females carried eggs that were large and visible through the lower half of each side of the abdomen. Windscorpion eggs normally become very noticeable days or weeks before they are laid, as late stage development occurs just beneath the transparent abdominal wall. After a week the first female laid eggs, but only those from one side of the abdomen. Eggs remaining in the abdomen liquefied, and the female died a few days later. The eggs that were laid collapsed at approximately the same time. The second female deposited about eighty eggs a few days later on August 7. These eggs did not collapse, and were removed the day they were laid. No eggs were visible in her abdomen, and it seemed she had completed her final duty of reproduction. About ten days later, eggs were again visible through the second female's abdominal wall. Five days later she laid a slightly smaller batch of eggs. She seemed sluggish (seemingly nearing the end of her life), and since windscorpions are often reported to guard eggs, she was left with them. On day three she returned to her normal energy levels, possibly, in part, because she had consumed all the eggs. After another two weeks she laid a third batch of eggs. These were removed. She lived till the end of September; very active one day and dead the next.

Galeodes Windscorpion

Ovum were maintained on a shallow bed of damp, somewhat dirty, sand. This was kept inside a 5oz condiment cup with three thumbtack holes in the lid. The temperature was maintained near 75°F. After a few weeks, most of the eggs began to show signs of development as dark and light differentiation of the yolk. Two dozen eggs never showed signs of development and a few of these became moldy over the following weeks.

The healthy eggs did not go through a post-embryo stage as commonly seen with common

FEMALE *EREMOBATES* WINDSCORPION

arachnids like tarantulas (inside the eggsac) and amblypygids (visible through the sac membrane). Postembryos look like an egg with an embryo sticking off the side, which has rudimentary development and no mobility. Instead, ova hatched into the first instar stage September 17-18. These resembled the first instars of scorpions, vinegaroons, amblypygids, etc. They were pale, slightly underdeveloped, but overall looked like later instars; a few did not hatch correctly and had twisted legs or abdomens. Many moved to the far side of the container, about 1.5", but after dispersal gave up all mobility. Nearly all laid on their backs, but did not touch the sand due to a dozen, barely visible, long, curved setae that kept them elevated nearly a sixteenth of an inch from the surface. The leg tips and chelicerae transformed to dark pink a few days before hatching. Fourteen days later, the young began to molt to second instar, and it took three days for all of them to molt. Molting may have taken longer than usual because the substrate had unknowingly been allowed to dry out. When moistened, the remaining handful of 1st instars molted within a few hours. The 2nd instars replaced the long abdominal setae with numerous, shorter hairs.

Behavior: Wild adults were collected at night as they came to lights to feed on insects attracted by parking lot lights. They'd run in the light circle, grab an insect and then run off. Healthy captives run around the cage, constantly digging and probing the surroundings. A small section of egg carton was offered as a hide, and it was inhabited now and then, but not used as a place to retreat to. At first, nesting behavior seemed to occur as the females would chew up egg carton and carry it to areas of the cage, but this material was not used during egg laying. Only 1/2" of substrate was provided, so females were unable to build tunnels or egg laying chambers. (Eggs were laid on the surface.)

GALEODES WINDSCORPION

Climbing: As usual, this windscorpion has no ability to grip smooth surfaces with its feet. However, it can still climb with amazing agility up the sides of the smooth walls of plastic and glass containers. Climbing appears very odd since only the two pedipalps (mouthparts) are used and the motion somewhat resembles a child swinging across monkey bars. As soon as the lid comes off, the females begin to swing up the sides. Specimens jump when they reach the top. By the second instar, immatures can climb readily, but it would be difficult to determine if first instars could climb since their period of activity is extremely short. While solifuges can use the sticky pedipalps to climb, other species I've kept have not done so with regularity.

Communal Immatures: The 1st instar immatures hardly move and have little choice but to remain together—the initial dispersal may be greater or

Eremobates Windscorpion Eggs

lesser inside a natural burrow, but is likely less than a few inches. Considering the incredible appetite of this predator, it seemed likely that shortly after molting to 2nd instar the immatures would begin to eat each other. However, the 2nd instars, which are capable of running, climbing, and feeding, actually huddle together tightly for the first few weeks. They regroup when disturbed. At three weeks they began to disperse and construct burrows, but didn't appear to cannibalize.

Eremobates Hatchlings Ready to Molt

They were offered springtails a few days after molting; some food may have been eaten, but the prey's jumping ability may have limited or prevented feeding. Flightless *Drosophila melanogaster* were offered at three weeks and were caught as they ran into the immatures, but no chasing was observed. Second instars were seen numerous times feeding in twos on a single fruit fly, but did not cooperate; each grabbed the fly from one end and proceeded to eat as quickly as possible. Considering the large relative size of the flies, sibling windscorpions would make excellent food, were size the only factor. Immatures that died were not eaten by siblings.

Eremobates Windscorpion Hatchlings

Prey Consumption: Immatures fed on early instar crickets and adult *Drosophila*. The opisthosomas of the young turn red after they've fed on a few fruit flies. They readily take prey as large as themselves. Unfortunately, many of the second instars showed no feeding response at three weeks, and died within a few days. If fruit flies or hatchling crickets were offered earlier, it's possible some mortalities could have been avoided, but a number that fed also died. It's possible a different type of prey such as firebrats could make a difference, but die-off doesn't seem to be related to feeding.

The adult females were given mostly small crickets, but were fed lobster roach, *Nauphoeta cinerea*, nymphs as well. Feeding frequency was approximately once every other day.

Remarks: Hopefully the information presented here will aid or inspire those interested in captive reproduction of this fascinating arachnid order. Early development may be different for other families in the windscorpion order (beyond simple timing). It's likely that all solifuges hatch directly from the egg into a first instar that cannot feed because the chelicerae are held apart and appear to be nonfunctional. The level of mobility and initial behavior may be different in some families.

Two and half months after the first eggs were laid, only three *Eremobates* survived, but few of the deaths seemed to have any cause. There is much to learn, and new discoveries covering the husbandry of this order are needed. It's unlikely that new information on captive husbandry will suddenly make rearing windscorpions easy, but specific husbandry experiences may allow for rearing of hardy species. Hopefully one of the amazing giant exotics that are imported in from time to time will prove to be such a species. (And, hopefully the keeper who acquires a specimen has the interest and information required).

EREMOBATES WINDSCORPIONS, 2ND INSTARS (RED BELLIES FROM EATING FRUIT FLIES)

EREMOBATES FEMALE, FROM CALIFORNIA

SCORPIONS

EMPEROR SCORPION
PANDINUS IMPERATOR

Display: 5- to 30-gallon cage with damp, shallow substrate and vented lid. Numerous specimens of the same size can be kept together. Scorpions fluoresce beautifully under black light, so a dark room or dark box can be used to display this feature. Cages should be secured from human access. The venom of this species is quite mild, but that means little to someone allergic to scorpion stings.

Background: Pandinus imperator occurs across west and central Africa. It is commonly imported for use in invertebrate displays and for hobbyists. Its prevalence in the trade is the reason it's listed on CITES Appendix II. This is one of the largest, bulkiest scorpions, which can be up to nine inches in length. However, most adult specimens are less than six inches.

Adults look similar, but can be sexed using the comb-like appendages on the underside. Males have 31 or more teeth on the pectines, while females have 30 or less.

EMPEROR SCORPION

EMPEROR SCORPION

Husbandry: Temperature affects health, but has a major impact on growth and reproduction. Usually adults are purchased for display, and they are kept in the low 70°s for long periods. These cool temperatures reduce activity and feeding, and lead to eventual failing health after a year or so. The scorpions are active and feed well closer to 85°F. Adults should be kept cool, around 72°F, for three to four months, and then raised up to 85°F for mating. Some keepers suggest the temperature also drop each night, but this probably is not necessary. If a pregnant female is kept cool for long periods, young will never be born.

During mating the adult pair will hold hands and dance, as the male positions the female to accept the spermatophore. As many as 40 scorpions will be born following a year of gestation. The pale white babies climb onto the mother's back, where they will stay for two weeks. At this time they molt and the mother can be removed from the cage.

FLAT ROCK SCORPIONS
HADOGENES SP.

Display: 2- to 10-gallon cage with a tiny water dish and a vented lid. Only one specimen is kept per cage. Scorpions fluoresce beautifully under black light, so a dark room or dark box can be used to display this feature. Cages should be secured from human access. The venom of this species is quite mild and it rarely stings, but an allergic response to stings is possible.

FLAT ROCK SCORPION (*HADOGENES PAUCIDENS*)

Background: Two species from Africa are commonly available: *Hadogenes troglodytes* and *H. paucidens*. The latter has been more common in recent years, and is easily differentiated by light banding on the abdomen. (It is often referred to as the "olive-keeled flat rock.") *Hadogenes troglodytes* is considerably larger than *H. paucidens*. Some males reach nine inches (including the long tail).

Adults are easily sexed by a quick look at the metasoma or "tail." The female has a tiny little tail that seems too small for the animal. The male's metasoma is longer than the body. Although extremely long, it is thin relative to the body of the scorpion. Due to the large size of the animal and this odd feature of the male, *H. troglodytes* is often listed as the longest scorpion.

Husbandry: A dry environment is preferred, but there should be a small water dish available. The dish should be very shallow and much too small for the scorpion to fit in and drown. *Hadogenes* like to hide under and between flat sheets of plastic or rockwork. Stacked flat pieces of slate (old slate shingles are great) or other flat stone with smaller stones glued underneath (as spacers to create a gap of approximately 1/2") make great hiding places. *Hadogenes* do not care for walking on typical loose substrate if they can avoid it.

Captive reproduction is uncommon. Gestation is long, about a year and a half, and results in the birth of two to three dozen young.

NORTHWESTERN FOREST SCORPION
UROCTONUS MORDAX

Display: 2- to 30-gallon cage with damp substrate and limited ventilation. Numerous specimens of the same size can be kept together. Scorpions fluoresce beautifully under black light, so a dark room or dark box can be used to display this feature. Caging should be secured from human access.

Background: Northwestern forest scorpions occur in California, Oregon and Washington, where they can be rather common in heavily forested areas around dead trees and rocks.

There used to be quite a few *Uroctonus* species, but *U. mordax* and two others are all that

NORTHWESTERN FOREST SCORPION WITH BABIES

remain. (Taxonomists moved the rest to other genera.) There are two subspecies, *U. m. mordax* and *U. m. pluridens*.

This scorpion is minimally defensive and known to have rather mild venom, but it should still be handled carefully to prevent stings.

Husbandry: Specimens should be provided high humidity and burrowing opportunities. These handsome creatures require minimal attention and few keepers have trouble maintaining them for years. Specimens are not very aggressive towards each other and are usually kept in groups, as long as there isn't extreme disparity in size.

Northwestern Forest Scorpion, Pregnant

Northwestern Forest Scorpion

This moderate-sized scorpion (1.5-2″) is one of the few species that actually does well at room temperature; the vast majority of scorpions require much higher temperatures for growth and reproduction. Since it is a temperate animal, some keepers refrigerate specimens during the winter. A cool period isn't necessary for general husbandry.

Females are notably stockier than males, making the adults easy to sex at a glance. Additionally, the male pectinal tooth count is usually a dozen to fifteen, while females average three or four fewer pectinal teeth. They take two years to reach adulthood, and then continue to live two or three times that long. Females have one to two dozen young after close to a year of gestation.

Up to two-dozen offspring are produced. Females with young should be removed from communal caging as other adults may eat immatures. She may be kept with her brood for a few months after they've become independent. She will share crickets she catches with them, which does seems to improve their survival rate. Remove her immediately if any young disappear, as all mothers are not equal.

CRUSTACEANS

Terrestrial Isopods

Display: Difficult to display due to small size and cryptic behavior. Useful for classroom observation and maze experiments (right, left, right avoidance).

Background: It could be argued that isopods are the most successful animals on earth. They are an ancient order that long ago conquered land and sea. They are common nearly everywhere, even though they aren't the largest group in terms of number of species. There are only 130+ terrestrial species in the U.S., but if a person walks outside and lifts a rock or log, isopods are likely to be the first creatures encountered.

Armadillidium nasatum

Porcellio laevis

Terrestrial species thrive in most habitats, from shore to desert. They have a dorsally-flattened, oval body and some can roll into a ball when disturbed. The legs don't stick out much past the sides of the body. The largest terrestrial species grow to 3/4", while the semi-aquatic littoral species reach 1.5".

They are so common in the ocean that three small species have found their way into my mini-reef aquarium (hitchhiking on "live rock"). Most marine species are so small they go unnoticed, but the largest species of isopod, found in the Gulf of Mexico and West Indies, grows to 16". Many marine species look similar to the familiar terrestrials, while others are greatly modified. Many are sexually dimorphic, with the male's rear legs, mandibles or uropods greatly enlarged. (Uropods are the two appendages at the end of the abdomen.) Some are parasitic on fish or other invertebrates. Parasites often go through different stages where the adults or young look nothing like the standard isopod. Other isopods have conquered the depths of the ocean and have been found more than 6 miles deep. There, colossal, pale-white isopods grow to the size of a football.

A number of species inhabit freshwater, but they aren't as dominant in this environment. The aquatic genus *Asellus* is available from biological supply houses and grows to half an inch. It looks more like a flat millipede than a traditional isopod.

At a glance, terrestrial isopods resemble centipedes or millipedes. Pillbugs and pill millipedes are often confused. However, isopods have a defined abdomen, body segmentation, appendage placement and other features that place them with our favorite seafood. (Class Crustacea, Subclass Malacostraca—isopods, shrimp, crabs and lobsters).

ORANGE *PORCELLIO*

Reproduction and Care: Isopods sexually mature long before attaining full size. This feature seems strange because many common arthropods don't reproduce until they are full-grown. However, this feature is the norm for crustaceans. Most terrestrial isopods begin to produce offspring four to six months after hatching. Adults continue to molt and grow and require a few years to reach full size. They eventually stop growing, but live at least three years, possibly much longer.

The genders of most terrestrial isopods appear identical from above. A female carrying eggs has a very large, obvious brood pouch or marsupium located between the legs. The following is

DALMATIAN *PORCELLIO*

detailed gender identification outlined by Dr. Thomas Carefoot:

"Males are identified by the presence of sperm transfer appendages located on the first two segments of the abdomen. It is necessary to look at them through a magnifier and use a very fine pin or wire to reflect up the appendages, which are usually held close to the body for protection. To identify a female, look for the presence of the oostegite appendages on the female's under-thorax, used to hold the eggs. There are seven pairs of these, and their presence is indicated initially by small nubs arranged symmetrically on either side of the thorax, just above (i.e., closer to the body) each walking appendage. Prior to her breeding molt the oostegites appear fully grown and are expanded to form the typical marsupium of a gravid (egg-bearing) female."

Maintain a 1" layer of dried or decaying leaves to encourage continuous reproduction. Terrestrial isopods consume all sorts of fruits and vegetables, dead inverts and any human or animal food. They grow well on most diets, but produce young less commonly or not at all without the leaves. Nearly any leaves are accepted, though a mix with oak, rose, viburnum or beech works well.

A good rearing cage has a screen or vented top and smooth sides such as the easy-to-find, plastic creature cages. Ventilation is important—primarily for pillbugs—but the substrate should never be allowed to dry out (except for the very surface). Isopods can't climb smooth, vertical

PORCELLIO WITH IRIDOVIRUS

surfaces. Place 2-3″ of potting soil or peat on the bottom, and don't forget the decaying leaves.

Isopod Color Varieties: There are some isolated color varieties of *Porcellio* sp. in culture. This species is common in the U.S., but originally came from Europe, possibly with the first settlers. The gray form is an excellent feeder because it reproduces rapidly and easily in captivity. The orange form is hardy and almost as fast breeding. Hatchlings are born orange and breed true if grays and "calicos" are removed for a few generations. Various related and unrelated isopods have orange and pale forms, but they tend to be wimpy and not brightly colored. The recent "Dalmation" variety is white with variable black flecks and black eyes. Sadly, the incredible electric blue form is very sensitive and is not born blue.

Isopods and their relatives can be infected with iridoviruses that changes the color of the exoskeleton. The "electric blue" crayfish encountered in pet shops is supposed to be brown, but is infected with this type of virus. Numerous isopod species can be infected, though many are immune or partly immune. The iridovirus has been successfully introduced to various unrelated isopods. The result is beautiful, but not harmless like it is to crayfish.

Isolated blues die in a few days to a few months. It's difficult to determine how quickly they die in a mixed culture. Isopods eat their own dead, which may be the only transmission method for this iridovirus. Blues of various ages show up in infected gray cultures, but any that survive long enough to reproduce have gray young. Blues can be very difficult to maintain, and quickly die out of cultures kept in substandard conditions. High moisture and available food helps maintain them. It might be possible to slowly remove grays over time to create a pure blue culture, but if the transmission method is as limited as it seems, that goal would be unobtainable.

ORANGE *PORCELLIO* WITH IRIDOVIRUS

Commonly Cultured Species: "Giant Canyon" *Porcellio laevis*—this large stock was collected in a canyon in California, though the species can be found in many states. Older specimens can grow to 4/5″ and are sexually dimorphic, as in older specimens the male uropods are twice the length of the female's. Developing specimens often turn white, but change back to gray after a molt or two if isolated.

"White micropod"—the original animals probably came from New York. Most isopods start out smaller than a fruit fly and begin pale to nearly white. At a glance these seem like immature isopods, but after two years the adults remain white and barely 1/8th of an inch. In addition, this species begins life smaller than a grain of salt.

"Jungle micropod"—this tiny species from Costa Rica is prolific. It is sensitive to cold temperatures. Adults do not grow as large as white micropods.

Pillbugs, *Armadillidium*—otherwise known as "pill isopods," two species are common throughout the U.S.: *A. vulgare* and *A. nasatum* (both introduced from Europe). Adults produce huge litters of young, but babies grow for a year or two before they are large enough to breed. *Armadillidium vulgare* grow larger, but reproduce and grow more slowly. Old adults can reach 5/8" but it takes two or three years.

"Red Claw" Shrimp
Macrobrachium sp.

Display: 5- to 10-gallon aquarium. One male and a few females can be kept together.

Background: Large freshwater prawn were a popular sight at local pet shops in the mid-1980s. They were handsome, huge animals which lived for years (often shrinking a little with each molt), but weren't bred in captivity due to the difficulty of rearing planktonic larvae. The ubiquitous tiny, translucent ghost or grass shrimp have been available as feeders for decades. In recent years a handful of colorful, miniature shrimp that breed easily in captivity have gained some popularity, while the much larger, wild-caught, filter-feeding wood shrimp (again planktonic larvae) have been common at pet shops for a few years.

This shrimp resembles the once-common giant prawns, but is a more manageable size for the aquarium, and in many ways is the perfect shrimp for small displays. *Macrobrachium* sp. "Red Claw" has been popular in the aquarium hobby in recent years due to ease of reproduction, appropriate size, and extreme hardiness. It is also much more accepting of tank mates than other macros. The number one allure of this macro is its ability to survive and thrive in captivity.

This particular crustacean from southeast Asia can be found listed under half a dozen different scientific names, of which none may be correct. *Macrobrachium* is a huge genus comprising more than two hundred species. Some of the largest species are farmed for human consumption, most notably the popular, foot-long, blue-clawed *M. rosenbergii*.

Macrobachrium with Eggs

Reproduction: The genus name refers to the macro, elongate arms of the male. Young males and females are difficult to tell apart because the front arms appear similar. Adults males grow to 3"—not including the long front arms—but are capable of mating at just under an inch. At this size they are similar in color and shape to an adult ghost shrimp. Females only grow to 2", but when they first mature they're the same size as males. This makes gender determination very difficult early in maturity. Males fight but can coexist. A hierarchy is established and only one male will grow to full size in a small aquarium. The larger the male, the larger the claws and deeper the color of his forelegs.

As long as there's a male and female, the female will carry a batch of eggs every two to three months. The male must mate with her shortly following her molt, so they should be kept together for breeding. If she's small (young), there

MALE *MACROBACHRIUM*

MARBLED CRAYFISH WITH EGGS

may be less than a dozen eggs, whereas large females can carry three dozen or more. The eggs start out bright green and change to a gold color a few days before hatching. Gestation requires around thirty days.

Hatchlings are 3/16" and are see-through with a bright red spot at the base of the claw.

Habitat and Food: This species is capable of growth and reproduction at 64° Fahrenheit, but 75-85°F is recommended for good development and gestation. Sexual maturity can be reached in four months, but, as usual for crustaceans, they will be nowhere near full size. A year or more is required for a male to obtain maximum size. Adults live two to three years, rarely longer.

Body color is affected by lighting, temperature, age and genetics. If kept with extremely limited ambient lighting, even very old, large males will be transparent with a few blue-colored markings (females will have even less color). Moderate ambient lighting or direct lighting causes darker coloration, though extremely bright lighting doesn't lead to even greater pigmentation. The male's front arms eventually darken to a rusty red-brown as they grow older. Under identical conditions some females will be clear and others will develop greenish, tiger stripes.

A few dozen shrimplings can be kept together safely in a ten-gallon aquarium, but as they grow larger they begin to cannibalize until fewer than ten are left. They are cannibalistic when crowded, and missing legs aren't rare. Cage surface area versus number and size of shrimp is the primary consideration. Extra feeding doesn't deter cannibalism, though starvation would likely increase cannibalism. Decorations like plants and floating wood adds to the livable space.

The choice of additional lifeforms in the aquarium is limited. Unlike other members of the genus, "Red Claw" macros seem to have lost the ability to catch live fish. The problem is many types of fish will damage or eat vulnerable shrimp following the molts. The shrimp love to eat snails and even consume the pesty Malaysian trumpet snail that's too well-protected and patient for other predators. (Snails greater than 1/2" are usually safe.) They pick at dead leaves but tend to leave live plant parts alone.

Food is offered 1-3 times daily. They eat any pellet or flake food designed for fish, but also dead leaves and any foods that won't foul the water. Some hobbyists add Indian almond leaves as a backup food, but these turn the water yellow, and aren't any better than free, old, brown elm leaves. Like fish, they learn to recognize the feeder and come running on approach. They are more than happy to be hand fed with a piece of frozen brine or krill.

An airstone or sponge filter is recommended to keep the water aerated since power filters can suck up the babies. Neither is required with minimal feeding and limited numbers, or when kept in an outdoor tub.

Conclusion: Macrobrachium sp. "Red Claw" is extremely hardy, breeds and grows without prompting, and is the perfect size for small aquaria, while superior to the popular, tiny species that require keen eyesight. It's the ultimate shrimp for easy captive care.

Marbled Crayfish
Procambarus sp.

Display: 1- to 10-gallon aquarium. Up to half a dozen adult specimens can be kept in a 10-gallon.

Background: Though it is a popular aquarium species kept around the world, this animal is a puzzle. It was "discovered" in Europe's aquarium hobby, but biology suggests it's an American species. Although closely allied to other *Procambarus* species in the southeastern U.S., it has never been encountered in the wild. It has no species name. Being discovered on the wrong continent is incredible, but what's even more puzzling is its mode of reproduction. It is the only known species of more than 15,000 decapod crustaceans

"ELECTRIC BLUE" *PROCAMBARUS* CRAYFISH (IRIDOVIRUS), WITH 2ND INSTARS

to reproduce through parthenogenesis. Males do not exist in culture and might not exist on our planet.

Decapods have ten legs, as the name suggests, and are usually considered the most advanced crustaceans. Until recently, evidence suggested a male and female were always required for reproduction. The most familiar decapod groups

OLD FEMALE MARBLED CRAYFISH

are crayfish, lobsters, crabs and shrimp. Crayfish are freshwater relatives of the marine clawed lobsters. (Slipper, spiney and other lobsters are more distant relatives of both.)

Crayfish are found across the globe in freshwater environments, and include more than five hundred species. Three hundred or so of these are found in the southeastern U.S., the hotbed of crayfish speciation. Some species can be fully mature at one inch, though most are over two inches. The largest species, *Astacopsis gouldi*, hails from Australia and grows to double the size of edible lobsters (the size normally encountered at the local grocer or restaurant).

Marbled crayfish are sexually mature at 2″ but can eventually grow to 3″. The exoskeleton is mottled in dark and light gray with tan, green, and blue highlights. Coloration is strictly influenced by the environment since the offspring are genetically identical. This species can turn blue if infected by the crustacean iridovirus.

Husbandry: Marbled crayfish can be kept in nearly any freshwater aquarium, at any temperature.

Filtration is great, but any container of water is usable. If overfeeding or bad care is expected, water should be shallow, so the crayfish can breath air.

Fish food flakes or pellets are the common foods given to marbled crayfish in captivity, but they can eat nearly any animal or vegetable matter. Hellweg claims this is the only animal that eats Java moss (*Taxiphyllum barberi*), which is only partly true; any crayfish eats the unpalatable plant. Java moss is usually kept with immatures to provide places for escape and hiding, but if adults are in the cage it can be consumed rapidly.

Marbled crayfish do eat each other, but are less prone to cannibalism than most crayfish, and are kept in groups with minimal losses.

The number of molts required to reach maturity is difficult to observe, but may be up to two dozen. Under decent conditions they'll go from hatchling to holding eggs in about eighteen months.

Vampire Crab
Geosesarma sp.

Display: 5- to 20-gallon cage with no lid. Specimens can be housed together, but there should only be one adult male per cage. Hiding areas or substrate should be avoided, or they remain out of view.

Young Vampire Crab Molting

Background: These brightly colored land crabs from Sulawesi offer a beautiful display and are truly a gem, as they have no planktonic stage and so can be bred in terraria. There are a few color varieties that may or may not be the same species, but here we'll cover just the original vampire. The legs and body are dark purple and the front claws are light purple. The top of the carapace is covered in a large gray splotch and the eyes are bright orange. Some say the eyes are the vampire "fangs," or the markings on the face look somewhat like a vampire. The origin of the common name is not obvious. Genders appear similar, but are easily sexed by the tail that is held up against the bottom of the carapace. It is very easy to see on adults and large immatures, and much less pronounced on very young specimens. The male's tail is narrow, while the female's is nearly as wide as the body. Be wary—"pairs" for sale are often two males. Immatures are cryptic and look like a little rock or piece of dirt.

Husbandry: The cage should have water and land areas that are easily accessed with a rough, sloping ramp. They can't climb smooth surfaces, which is why a lid is unnecessary. The adults can usually access a shallow dish (such as the lid of a 1oz cup), but the young are too small. This species is considered terrestrial, and is capable of molting on land, but prefers to molt underwater. One option is to cut 2" thick sponge, and keep the water level below the top of the sponge. A thin layer of dirt is placed on the sponge and moss (including Java moss, which is very hardy and doesn't need to be submerged) or liverworts can be grown on top. The size and shape of the water areas is easily adjusted to individual preferences. Adults tend to stay on top of the sponge, while young crabs are often on the sides just below the water line if there's plant cover.

Temperature and humidity requirements are somewhat controversial. 85°F is often recommended as optimum, but I reared up a pair from hatchling and successfully produced young in the low 70°s without any molting or developmental

FEMALE VAMPIRE CRAB (TOSSING YOUNG FROM UNDER HER TAIL)

problems. Development was quick, less than five months. High humidity is often recommended, but terraria without adequate air movement quickly suffocate the crabs.

The primary diet should be made up of fish food pellets (those made primarily from fish and shrimp meal). Small crickets are a great treat for older specimens. They will feed on a variety of animal and vegetable matter.

Adults are often kept in groups but males are known to fight and will tear each other's legs off under indeterminate conditions (sometimes they get along without incident). Reproduction information is limited, but females seem to wait a full year after reaching maturity before producing. Like shrimp, crayfish, and other crabs, the female holds her eggs under the tail, but development can't be observed. The female's tail is raised a fraction of an inch and pleopods cover the edges, so it's easy to notice, but eggs and young are not visible. Do not handle a carrying female, or she'll reach in and start flinging out nauplii with her claws. She will not regather them and they can be weak and have high die-off, depending on how premature the separation. They are free-living at just under 2mm across the body, and are white until they molt. Adults eat any immatures they can get their claws on, so young should be moved to their own cage, if much of a survival rate is hoped for. Nevertheless you may find a missed immature months later, hiding in the substrate. Young crabs can be reared together with minimal losses, but as they grow larger, legs will end up missing if overcrowded. They molt every few weeks and even stragglers mature in six months.

TRIOPS

Display: Most clear plastic containers can be used to display and house specimens. Healthy animals swim in continuous circles and loops and make great, though short-lived, displays. New eggs must be hatched every three to five weeks to stock a continuous display.

Background: The popular and generic name *Triops* means "three eyes"—so named because they have a third naupliar eye near the front of the head between the two eyes. These peculiar crustaceans are also known as tadpole shrimp, since specimens are most often found in temporary pools and puddles. In murky water the outline is easy to mistake for a tadpole. (*Triops*, however, eat tadpoles.) *Triops* are commonly called living fossils since one species is the oldest known living species on the planet.

The most conspicuous body part is the large, oval carapace that resembles the cephalothorax of the chelicerate horseshoe crabs. However, the carapace is part of the head and is not fused with any thoracic segments. Due to a superficial resemblance, *Triops cancriformis* is often listed under the common name "horseshoe crab," but it could hardly be more different and it's not even in the same subphylum. The head covers most of the body except for the long tail, and its elongate, whiplike arms. The arms are the most notable of the thoracic segments known as thoracopods.

Most familiar crustaceans, including lobsters, true shrimp, crabs and isopods, belong to the class Malacostraca. *Triops* are crustaceans in the class Branchiopoda, along with the famous brine shrimp ("sea monkeys"), clam shrimp, fairy shrimp and daphnia. The *Triops* family includes two genera whose appearance is nearly identical, *Triops* and *Lepidurus*. There are a number of *Triops* species kept by hobbyists, including *T. australiensis, T. cancriformis, T. granarius* and *T. longicaudatus*.

Triops are quite large for temporary pool inhabitants, and they are most often between the size of a dime and a quarter at maturity. The adult size and length for *Triops* is a fluid concept, compared to most animals. As common for crustaceans, they continue to grow significantly after obtaining sexual maturity. Ultimate size is greatly affected by the size of the cage, and can also vary by individual.

Most available measurements for *Triops* include the carapace, abdomen and long segmented terminal appendages called caudal furca. Carapace measurement alone will be used whenever possible here. The comparative length of the appendages to the body is different according to species, genders and ages, so the carapace measurement is more representative of the size of the animal.

TRIOPS CANCRIFORMIS

TRIOPS LONGICAUDATUS

Reproductive Strategies: Triops granarius only reproduces sexually, so both a male and female are required. Other species may reproduce sexually in some populations, but usually reproduce through parthenogenesis, or are hermaphroditic. Hermaphrodites can either mate or reproduce through parthenogenesis.

Eggs and Hatching: Each ovum is incredibly tiny, not much bigger than a human hair in diameter. (For those familiar, *Triops* ova are slightly larger than brine shrimp or fairy shrimp eggs.) Bright red ova are glued into the substrate in egg masses normally containing 10-40 eggs. A female can construct an egg mass each day, so a long-lived female can produce thousands of eggs.

Eggs are very resistant to drying and heat, and can hatch even after a decade or two. Temperatures below 20°F and above 100°F have little effect on dried eggs. Many of the eggs will hatch after a single dry period, but some will not hatch for two, three or even more cycles. This helps them avoid extinction if conditions are bad one year. Egg substrate can be dried and retried later if hatching or rearing is unsuccessful.

Underside of *Triops cancriformis*

While drying periods are important, hatching also takes cues from dissolved solids and temperature. Distilled or deionized water is normally used for hatching because the higher dissolved solids in tap water can prevent some or many of the eggs from hatching. Other types of bottled water have been used with reasonable success, but some spring water may contain high enough dissolved solids to suppress all hatching. Temperatures in the 72-77°F range work well for common species. Higher and lower temperatures can work, but the further away the less likely eggs will hatch.

Since *Triops* cannot colonize new ponds on their own, the small size and extreme hardiness of the eggs plays an important role in distribution. Dirt or mud that sticks to a visiting bird's feet contains many eggs that can start a new population if some of the mud falls off when the bird visits another pond. Long ago it may have been flying dinosaurs or maybe even gigantic dragonflies transporting eggs in this manner.

Purchased eggs are ready to be used, but eggs produced in captivity need to be dried out before hatching is attempted. Many sources suggest eggs will hatch after three weeks of drying but three months is a more realistic minimum. (Possibly a tiny fraction of a percent of eggs could hatch after just three weeks of drying, under rare circumstances.)

Dried eggs eventually change from red to brown over a period of a few months. Red eggs will not hatch, but even after eggs turn brown they may still require more drying time.

When hatching eggs, keep in mind that only a few cups of water should be used with a teaspoon of egg substrate. If a large volume of water is used, the *Triops* invariably die. Microorganisms in the egg substrate condition the water, but this is prevented if a high ratio of water is used. They can be hatched in 16oz deli-cups filled 1/2-3/4, or a smallest-size, plastic "creature cage" can be filled to 1.5" depth.

Eggs begin to hatch in sixteen hours and may continue for a few days afterward. Eggs that do not hatch in forty-eight hours almost never hatch unless dried out again for a few months. If none of the eggs hatch because of high dissolved solids, a complete change to distilled water can cause eggs to start hatching as if they were just placed in water.

Growth: Even the "slow-growing" species reach maturity in a few weeks. Rapid growth is one of the most fascinating aspects of *Triops*. Unfortunately, it's followed by a rapid death. Habitat and genetics play a role, but the vast majority of *Triops* die from old age within 30-60 days of hatching.

Larvae (nauplii) hatch out barely larger than the egg and look like a diamond with two tiny arms. They aren't very mobile and remain in one spot moving their appendages and swaying back and forth. They molt a few times in the next twenty-four hours and begin to swim, resembling tiny adults. They continue to molt at a rapid pace, and go through at least a dozen instars before reaching 1/4″. At this point they can be moved to a larger cage. (If moved earlier, they nearly always die.) Adults continue to grow and molt every few days.

Maximum size is influenced by many factors, including genetics, cage size and food. If kept in the original 3.5″ round deli-cup, they will not grow nearly as large as when moved to a 8″ x 12″ container filled 3″ high. Pelletized food for predatory fish is likely as complete a diet as possible. If overfed, poor water quality will keep them from growing to their potential.

Feeding: Triops are predators that feed on copepods, seed shrimp, daphnia, fairy shrimp, tadpoles, insects, etc.. They are the ruler of the food chain in their preferred habitat. They are not obligatory carnivores, but are omnivores. In captivity, they can be fed a variety of foods, including most pet foods, fruits and vegetables. Despite what they *can* eat, the recommended foods include *Triops* food, *Betta* pellets and most granular fish foods.

For the first two days after hatching, they should not be fed. They feed on stored reserves and microorganisms in the egg substrate. (Eggs are always sold along with the sand or organic substrate.) At two days they can be fed a portion of one tiny fish or a *Triops* pellet smashed into barely visible dust. (Use pellets, as flake foods rot and grow mold too quickly.) Food should be entirely consumed in a few hours. As they grow larger they eat more, and adults will eat a few whole pellets each day. Different foods like pieces of dried krill and various vegetables shouldn't be attempted before the carapace is at least 1/2″.

Water Quality: Chlorine is found in tap water as a disinfecting agent and rapidly kills *Triops*. Chlorine dissipates if an air diffuser (bubbler) is placed in a bucket of water for forty-eight hours, or it sits for four days.

FEMALE *TRIOPS GRANARIUS* WITH EGG POUCHES

MALE *TRIOPS GRANARIUS*

Water changes can kill *Triops* if too much water is changed at one time. Remove only about 10% and repeat a few hours later if necessary. It is helpful to have a turkey baster on hand to remove detritus from the bottom of the cage, rather than removing surface water. The turkey baster can also remove uneaten food if too much was offered.

Triops can live in rather dirty water, but organic contamination, reduced oxygen and low pH can kill them or prevent adults from growing to full size. Once the animals are 1/2″ or so in diameter, an air bubbler or sponge filter may be used in *Triops* cages to improve water quality. Feeding the proper amount and small water changes every other day can be more helpful than a filter.

Though of minor concern and depending on when the coils were last cleaned, distilled (not deionized) water may pose an aluminum threat.

Husbandry Highlights: Use deionized water for hatching and water changes. In many areas tap can work fine, but be certain to let it sit at least 48 hours, with an air diffuser, to remove chlorine.

Moving animals or changing large amounts of water will kill small immatures and can kill adults. Do not feed for the first 48 hours, and do not overfeed at any time.

Triops cancriformis—this shrimp is known to have existed for 220 million years and is the oldest living animal species on the planet. This may be the ancestor of all other tadpole shrimp species. *Triops cancriformis* is one of the most common and widespread species in the world. Nevertheless, in the United Kingdom the local population is protected because it is only found in a single temporary pool. This is a handsome species with a speckled carapace.

When the carapace is just over .5″, brood pouches develop and become filled with bright red eggs signifying maturity—though they commonly do not lay eggs at this size. Brood pouches are constantly filled with eggs. Each adult can produce a few thousand eggs. Sexually mature specimens continue to grow until the carapace is 1.25″ long. In captivity, under similar conditions, older adults are twice the mass of *T. longicaudatus.*

Triops cancriformis are among the more mellow members of their family. Older individuals often sit in one spot on the substrate, whereas other species tend to be in constant motion. They barely react when bumping other individuals. Even in a small deli-cup they can coexist with other individuals without cannibalism. They eat most small fish food pellets and nibble on freeze-dried krill.

Compared to other common *Triops*, these grow slow and are long lived. They can be less than 1/8″ after a week. They commonly live 60-70 days from hatching.

Triops granarius—or "Japanese *Triops*", is one of the most widespread species and is often found in rice fields worldwide, along with *T. longicaudatus*. This creature is very different from other commonly kept species in that there are distinct sexes. The male tends to be much larger than the female and is easy to differentiate by the lack of brood pouches.

The female's carapace reaches about .5″, while the male is far more massive—his carapace reaches .75″. Females often have no eggs in the brood pouch, but if a male is placed in the cage the pouch fills with eggs within a few minutes. They can lay as little as a few hundred eggs.

Triops granarius are highly energetic and rarely stay still. When they run into a tank mate both recoil rapidly. Young are highly cannibalistic, so it is helpful to separate them. Full-grown animals can damage each other, but are not likely to cannibalize. They do well on small fish food pellets, but feed on dried krill with zeal.

Triops longicaudatus—if you've kept *Triops*, it's nearly certain this is the species you had. This species is commonly sold in "living fossil" kits (though it's not really a living fossil like *T. cancriformis*). It comes from the southwestern United States, but has been introduced in rice fields worldwide. In its natural range, different geographic populations can be bisexual, parthenogenic or hermaphroditic. Young are cannibalistic. This species is very fast growing and can die from old age in fewer than 30 days from hatching.

GASTROPODS

BANANA SLUG
ARIOLIMAX COLUMBIANUS

Display: Clear display caging can be difficult, since animals leave heavy frass and slime trails anywhere. Moving animals are impressive, but during the day they usually contract into a small ball with no physical features visible.

Background: Without a doubt, the banana slug is the most famous and sought after of the world's slugs. Its giant size—the second largest of all slugs—combined with brilliant color, make it the most impressive slug. Although an adult can stretch its body even longer than a grocery store banana, the slug is nowhere near the mass of the fruit. Many specimens are a vivid, banana-yellow and some have black spots—reminding one of cows and Gateway boxes. In some areas they are greenish and not so attractive.

Ariolimax columbianus inhabits the West Coast of North America from southern British Columbia to Northern California. A number of different species names had been awarded to geographical color varieties over the years, but most slug experts agree there are only a few closely related species, which are difficult to discern by anyone other than a taxonomist specialized in pulmonates.

Ariolimax have a lung and are hermaphroditic like other pulmonates, which include all terrestrial slugs and snails and most freshwater snails.

This incredible slug is extremely common in its natural range, but is rarely used in displays.

Husbandry: Banana slugs are notoriously fragile and they are often rapidly killed in captivity. Novice keepers often try to maintain the slugs at temperatures above 75°F, which immediately transforms the poor animals into lifeless piles of slime. Those aware of heat sensitivity may use the refrigerator to keep the slugs "alive" a little longer. A refrigerator is far too cold, so they hardly move and die in a few months or less. 62°-68°F is difficult to maintain without equipment, but at this range they behave normally, eat and seem to do very well. They enjoy strawberries, romaine lettuce and anything normal garden

YELLOW BANANA SLUG

SPOTTED BANANA SLUG

slugs and snails will eat. Unfortunately, even at these temperatures, they only live 6-12 months, and haven't reproduced in captivity.

A noteworthy feature of this particular slug is the thick slime trail that seems impossible to wipe or wash off. This slime detracts from cage aesthetics and can make handling undesirable. The more a keeper tries to wash it off with water, the slimier it seems to get. Slime can be dried and scraped off of caging but that's not realistic with the animal in the cage.

BROWN-LIPPED GARDEN SNAILS

BROWN-LIPPED GARDEN SNAIL
CEPEA NEMORALIS

Display: Clear display caging should be small to show off the animals, and quite a few can be kept together. They leave some frass and slime trails, but rarely enough to obscure viewing. They often rest in their shells, but are easy to revive with handling or misting.

Background: The name at first sounds terrible for such a beautifully striped creature. However, mature specimens form a thick lip at the end of the shell which is brown, and this feature usually makes it easy to differentiate it from a closely related species with a white lip.

BABY SNAILS

BROWN-LIPPED GARDEN SNAIL

Cepea nemoralis grows "wild" in most of our states. However, it seems to be adapted for developed and urban areas rather than wilderness. It was introduced from Europe. There are many different large pulmonate snails found in different areas of the U.S., some much larger than *Cepea*, but this species is great for displays because it has a colorful shell and it's active, hardy, and probably available in your state.

The brown-lipped garden snail was supposedly purposefully introduced for human consumption, but it's a bit undersized at only around 3/4" at maturity. The shells are usually ivory with a number of dark brown bands, but single banded and solid variants can be found in some populations.

Husbandry: This handsome snail is adapted to a wide range of temperatures, and does well in extremes far outside the comfort range of humans.

Specimens have a small appetite. They feed on mini-carrots, romaine lettuce, dry dog food and anything normal garden slugs and snails will eat. Depending on the calcium content of the food, it

A Terrestrial Snail from Ohio

may be necessary to provide cuttlebone or calcium sand as a supplement. If they lack calcium, they will have difficulty forming strong shells and will chew on tankmates' shells. Chewing removes the colorful surface and can cause death in extreme cases, as holes are chewed all the way through.

A breeding habitat should have a vented lid, two to three inches of coconut fiber, potting soil or other substrate, and a partly buried, sponge-filled water dish at one end. The sponge should be one solid piece and fill the entire container to keep out frass and waste, while maintaining a moisture and humidity reservoir. Water can be poured over the dish and surrounding substrate whenever substrate begins to dry. A few dozen to a hundred eggs are laid in a hole dug in the substrate, which is excavated in damp dirt near the sponge-filled water dish. Young snails from the same clutch of eggs can develop at notably staggered rates. Some immatures reach the size of a small marble within a month or two of hatching, while others from the same group can be little bigger than the egg six months later.

Brown-Lipped Garden Snails

BIBLIOGRAPHY

Arnett Jr., Ross (1993) *American Insects: A Handbook of the Insects of America North of Mexico*. Gainesville, FL. The Sandhill Crane Press.

Bíly, Svatopluk. (1990) *A Colour Guide to Beetles*. London: Treasure Press.

Boucher, Phyllis. (2001) *Crayfish*. Minneapolis, MN: Lerner Publications Company.

Castner, James. (2000) *Amazon Insects*. Gainesville, FL: Feline Press.

Dixey, Steven. (2002) Tanzanian giant banded tailless whipscorpions. *Invertebrates-Magazine* (Elytra and Antenna). (June 02).

Dunn, Gary A., ed. (1993) *Caring for Insect Livestock: An Insect Rearing Manual*. Special Publication No. 8. Lansing, MI: YES, Inc.

Eisner, T. (1958) Spray mechanism of the cockroach *Diploptera punctata*. *Science* (18 July): 148-149.

Evans, Arthur V., Bellamy, Charles L., and Lisa Charles Watson. (1996) *An Inordinate Fondness for Beetles*. New York: Henry Holt and Co.

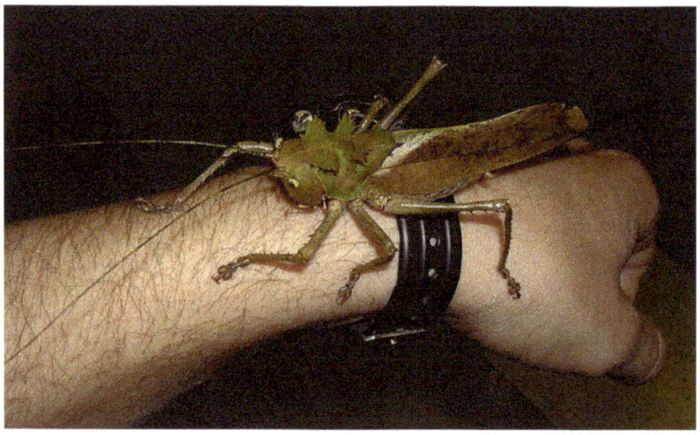

Malaysian Katydid, at an Insect Zoo

Fritzsche, Ingo. (2003) *Lucihormetica grossei* n. sp. -eine neue Art der "Leuchtschaben" aus Venezuela (Dictyoptera: Blaberinae: Brachycolini). *Arthropoda NR*. 11(4-2003): 7-13.

Helfer, Jacques R. (1987) *How to Know the Grasshoppers, Crickets, Cockroaches and Their Allies*. New York: Dover Publications.

Hellweg, Michael R. (2009) *Raising Live Foods*. Neptune City, NJ: TFH Publications.

Hetrick, L. A. (1949) A. Field notes on a color variant of the two-striped walkingstick, *Anisomorpha buprestoides*, (Stoll). *Florida Entomologist* 32(2, June): 74-77.

A Hands-On Beetle (*Eleodes*)

Hillyard, Paul. (1994) *The Book of the Spider*. New York: Random House.

Hoffman, Richard L. (1999) *Checklist of the Millipeds of North and Middle America*. Special Publication No. 8. Virginia Museum of Natural History.

McGavin, George C. (1993) *Bugs of the World*. London: Blandford.

McGavin, George C. (2000) *Insects, Spiders and Other Terrestrial Arthropods*. New York: Dorling Kindersley.

McMonigle, Orin. (2001) *Assassins, Waterscorpions & Other True Bugs Care and Culture*. Brunswick, Ohio: Elytra and Antenna.

McMonigle, Orin. (2003) *Giant Centipedes: The Enthusiast's Handbook*. Brunswick, OH: Elytra and Antenna.

McMonigle, O., and R. Willis. (2001) *Allpet Roaches Care and Identification Handbook for the Pet and Feeder Cockroaches*. Brunswick, OH: Elytra and Antenna.

Mizunuma, Tetsuo (1999) *Giant Beetles Euchirinae Dynastinae*. Endless Collection Series, Vol. 3. Tokyo, Japan: Endless Science Information.

ROACH PETTING ZOO

National Geographic. (2002) *Beetle Battles: Kwang Bang* (Gladiator Beetles, National Geographic Series: Nature's Nightmares). Australia: Becker Entertainment.

Papp, Charles S. (1984) *Introduction to North American Beetles*. Sacramento, CA: Entomography Publ.

Preston-Mafham, Ken. (1992) *Grasshoppers and Mantids of the World*. London: Blandford.

Preston-Mafham, Ken. (2000) *The Natural World of Bugs and Insects*. London: PRC Publishing.

Preston-Mafham, Rod, and Ken Preston-Mafham. (1993) *Spiders of the World*. London: Cassell PLC.

Reich, Joe. (2009) Keeping *Macrobrachium* sp. "red claw" shrimp. *Tank Topics (Greater Akron Aquarium Society)* (February/March).

Roth, L., and E. Willis. (1957) The Biology of *Panchlora nivea*, with observations on the eggs of other Blattaria. *Transactions of the A.E.S.* 83(4).

Rowland, J. Mark. (2003) Male horn dimorphism, phylogeny and systematics of rhinoceros beetles of the genus *Xylotrupes* (Scarabaeidae: Coleoptera). *Australian Journal of Zoology* 51: 213-258.

Sreng, L. (1993) Cockroach mating behaviors, sex pheromones, and abdominal glands (Dictyoptera: Blaberidae). *Journal of Insect Behavior* 6: 715-735.

Tait, N., and R. Vogt. (2005) *The Encyclopedia of Reptiles, Amphibians & Invertebrates*. San Francisco, CA: Weldon Owen.

Tydeman, C. (1982) Phylum Arthropoda. *Encyclopedia of Marine Invertebrates*. Neptune City, NJ: TFH Publications.

Walls, Jerry G. (1998) *The Guide to Owning Millipedes and Centipedes*. Neptune City, NJ: TFH Publications.

Werner, F. G., and C. A. Olson. (1994) *Insects of the Southwest*. Tuscon, AZ: Fisher Books.

Weygoldt, Peter. (1969) *The Biology of Pseudoscorpions*. Harvard Books in Biology, Number 6. Cambridge, MA: Harvard University Press.

AN INVERTEBRATE EXHIBITION

Weygoldt, Peter. (2000) *Whip Spiders (Chelicerata: Amblypygi): Their Biology, Morphology and Systematics*. Denmark; Apollo Books.

Wyttenbach, R., and Eisner, T. (2001) Use of defensive glands during mating in a cockroach (*Diploptera punctata*). *Chemoecology* 11: 25-28.

Zompro, O., and I. Fritzsche. (1999) *Lucihormetica fenestrata* n. gen. n. sp. the first record of luminescence in an orthopteroid insect (Dictyoptera: Blaberidae: Blaberinae: Brachycolini). *Amazoniana* XV (3/4).

www.ingramcontent.com/pod-product-compliance
Ingram Content Group UK Ltd.
Pitfield, Milton Keynes, MK11 3LW, UK
UKHW060224061026
15625UKWH00008B/491